FACTORIAL DESIGN

Thomas Elser

© 2017 by Thomas Elser

Title:	Factorial Design
Author/Publisher:	Thomas Elser, D-84556 Kastl, info@elserth.de
Layout:	Thomas Elser
Cover design:	Markus Käßler, D-84489 Burghausen, kaesslerm@freenet.de
English translation:	TechniText Translations, D-28857 Syke, info.technitext@ewe.net
Printed by:	CreateSpace, 4900 Lacross Rd, North Charleston, SC 29406, USA
ISBN-13:	978-1542906111 (CreateSpace-Assigned)
ISBN-10:	1542906113

FACTORIAL DESIGN

Understanding Design of Experiments (DoE)
and Applying it in Practice

About the author

After completing his engineering degree, Thomas Elser worked in the automobile and chemical industries on statistics, factorial design and quality management, among other things.

His approach to the topic discussed in this book resulted from his many years of practical industrial experience as a Six Sigma Black Belt in research and engineering and his work as a lecturer of courses for engineers/technicians. Realistic tasks offer an easily understandable introduction to factorial design. The objective is to provide the reader with the confidence to apply and evaluate factorial designs at the practical level, and particularly to enable them to use the appropriate software professionally and successfully.

Content

1   Design of Experiments – Fields of Application and Objectives

The optimisation of systems and processes is an enduring topic in research, development and manufacturing in all sectors of industry. Improvements can be made to specific performance characteristics of technical installations and machines. Or more generally: physical/technical, chemical, medical or business processes etc. can be optimised. On the practical level, several obstacles need to be overcome as it is seldom the case that the workings of the system under consideration are known in sufficient detail to enable a mathematical simulation to be undertaken. Practitioners struggle with compromises when not all the operating parameters of the process to be optimised are known, when their number is very large and/or when the strength of their effects is unknown, for example.

The task for an experimenter could be, for example: a series of experiments have to be conducted to determine the factors affecting the gas consumption of a turbine. When developing rechargeable batteries for electric cars, the task could be to determine and quantify the factors affecting the charging time. The task at hand can come from process technology as well, however: what are the effects of boiler temperature and pressure on the yield of an active pharmaceutical ingredient in a pharmaceutical reactor, for example?

There is more to it, however. You will see that Design of Experiments is not limited to technical systems and processes. There are also a great many disciplines with processes where the effects of various parameters on specific responses are of interest. Consider, for example, the extensive field of surveys/interviews related to sociological or medical/psychological processes. What effect does the weather on election day have on the behaviour of voters in particular age groups? Or: what is the effect of the dose and the time a medication is taken on a particular blood test result?
Business and logistics sequences are also processes and have input variables and responses. One task could be to use a factorial design to investigate which factors affect the delivery times of specific products and how great these effects are.

In general terms, Design of Experiments is concerned with identifying input parameters of processes, changing them in a systematic way, and measuring and statistically evaluating their effect on a process response. The objective is to optimise the process in respect of certain responses. Particular attention must often be placed on the economic aspect: Design of Experiments claims to achieve results with a very high degree of reproducibility for as little experimental effort as possible (= small number of expensive experiments).

The standardised procedure for the Design of Experiments method has meanwhile been accepted all over the world. And the standardised representation of the results forms the basis of efficient communication in this field. In English-speaking countries, in particular, this method is widespread in practice and an integral part of many university courses.

Adopting an analytical approach using simulation software does not work with many processes because no mathematical models are available for them. This is where the Design of Experiments method is of great practical importance since it views the process as a closed system ("black box"). It is not interested in its internal relationships. Instead, its approach is to systematically change the input variables (also called factors or parameters), and measure and statistically evaluate the effect this has on specific responses (see Figure 1).

The input variables are usually called $x_1, x_2, x_3 \ldots x_i$, the response is y. The objective is to utilise factorial design to obtain a prediction model $y = f(x_1, x_2, x_3 \ldots x_i)$ for the behaviour of the response for all combinations of the factor settings.

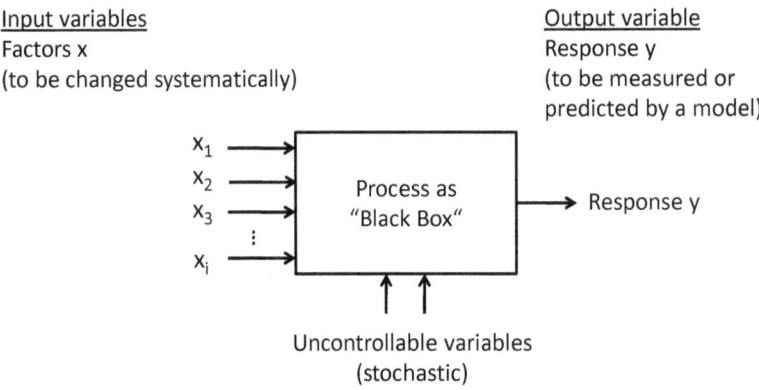

Figure 1: "Black Box": the inner workings of the process are not considered

In addition to the input variables, uncontrollable variables which are unknown and/or unavoidable can also affect the process. In many cases, they are the natural variances of the processes and the measurement systems used. Uncontrollable variables can also conjure up effects which do not even exist. An important issue in the mathematics of factorial design is to assess the so-called significance of the effects of the input variables on the responses, i.e. to distinguish real effects from those which can be explained by the uncontrollable variables.

A process usually has not only one, but several responses. A separate design must be drawn up for each one. Responses can be physical characteristics which can be measured directly or indirectly. In practice, derived quantities such as quality characteristics, for example, are also responses of designs.

The approach used in the book and the target group

A great many IT tools are available for designing and evaluating experiments (see Chapter 10). The calculation examples reproduced here were generated using the Microsoft Excel® and OpenOffice Calc® spreadsheet programs. The calculations and graphs on which the tables are based form the arithmetic of the Design of Experiments method, which is actually quite simple. This ensures that the numerical data contained in the book can be understood by "non-mathematicians" as well. Moreover, special software is also available for the design of experiments, which can also be used to realise more complex plans and calculations with ease. The graphical representation of the results is provided at the same time to make things easy.

The book provides a step-by-step introduction to the Design of Experiments method with the aid of simple mathematics (linear algebra covered in middle school). It uses simple, easily comprehensible examples with a manageable quantity of numerical data. The first objective is that the reader understands the underlying principle. This provides them with the confidence to apply the necessary IT tools correctly and be able to assess their results.

Scientists and engineers are initially introduced to experiments and how to plan them. These professions naturally form a large group of users. In addition, there are further issues relating to the design of experiments, in the professional groups of physicians, psychologists, social scientists for example, as mentioned above.

The book is written for students of all disciplines who have to or choose to tackle the topic as part of the curriculum of their course of study. It is also a reference work for the large group of users working in research, development and manufacturing in many different sectors of industry when they come to the practical application of their work.

The next chapter will show you that the systematic approach is very important in the design of experiments, because if this is lacking or insufficient, there is a danger that incorrect conclusions are drawn. Furthermore, there is a risk of wasting expensive experimental resources, such as staff/time, equipment and with destructive tests, in particular, the destruction of expensive products. Examples here are crash tests for cars or strength tests with expensive components.

1.1   Advantages of the Design of Experiments method compared to other methods

"The more, the better"

This catchword can be used to describe a widely utilised approach to the planning of experiments: the first factor remains set to one value, while the second factor is gradually increased after each experiment. The first factor is subsequently increased by one step and then the second factor step-by-step, and so on. This produces a more or less fine grid of experimental results depending on the size of the steps. The procedure appears to be logical and productive. But do the results always justify the considerable effort which may be necessary?

This method of "the more, the better" is now explained with the aid of an example (Figure 2): the reactor of a pharmaceutical manufacturer should supply a certain quantity M of an active ingredient per hour. It is known that the product quantity M depends mainly on the temperature T and pressure p at which the process is run. The plan is to increase the temperature of the reactor step-by-step for three pressure levels p_1, p_2 and p_3, and to measure the quantities of active ingredient produced.

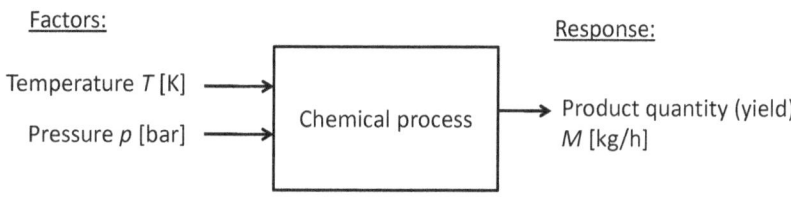

Figure 2: Black Box: product quantity (yield) of a chemical process

The disadvantage of this approach can be the considerable experimental effort needed: for each desired pair of temperature and pressure values, one experiment has to be run and the product yield has to be measured.

Let us assume the process itself takes four hours and the number of temperature levels to be run from T_1 to T_n is 20. Then, according to Table 1, 3 times 20 = 60 experiments have to be done for the three pressure levels assumed, and the duration of the experiments alone would amount to ten days and nights. Set-up times for the cleaning, charging and adjustment of the installation between the experiments would have to be added to this. With the aid of this example, it is easy to estimate the length of time and the high costs that this procedure would require. Whether this system would produce the desired information about the process is examined in a little more detail in the following.

	T_1	T_2	T_3	...	T_n
p_1	1	2	3	...	n
p_2	2n+1	2n+2	2n+3	...	2n
P_3	3n+1	3n+2	3n+3	...	3n

Table 1: 60 experiments have to be conducted for n=20 temperature levels and 3 pressure levels

The OFAT method is a popular choice - to draw the wrong conclusions?

We would like to warn against a further method which appears at first sight to provide information about the process while using a smaller number of experiments.

The OFAT method (*One Factor At a Time*) is often utilised when optimising processes in order to find a minimum or maximum of a response, for example. It has a crucial disadvantage compared to the Design of Experiments method: with OFAT, one factor is varied initially after the first experiment and the response obtained is measured. The "better" of the two settings of this factor is maintained for the subsequent experiments and the next factor is varied, and so on.

This principle is explained with the aid of an example. The task is to minimise the fuel consumption of a passenger car in a specific load range. It is assumed that the three factors speed, tyre pressure and the octane rating of the fuel are essentially responsible for the amount of fuel consumed (see Figure 3).

Factors: Response:

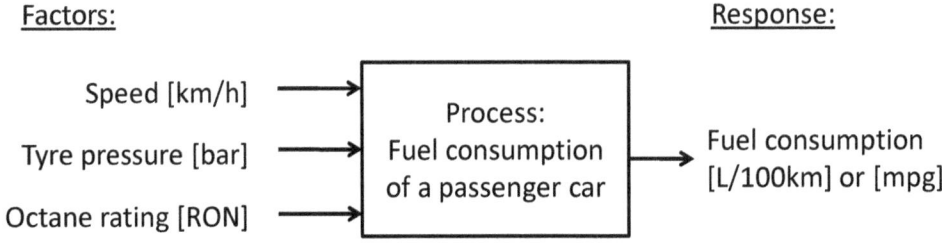

Figure 3: Black Box: fuel consumption of a passenger car

Table 2 contains hypothetical numerical values for the factors stated and the resulting response, which are used to explain the OFAT method. It is assumed that the three factors are each set to two levels.

Trial number	Factor 1 Speed [km/h]	Factor 2 Tyre pressure [bar]	Factor 3 Octane rating [RON]	Result (response): Fuel consumption [L/100km]	[mpg]
1	80	2.2	96	8.4	33.6
2	100	2.2	96	9.8	28.8
3	80	2.4	96	8.2	34.4
4	80	2.4	94	8.8	32.0

Table 2: The OFAT method: is trial no. 3 really the optimum?

12

The method is surprisingly simple and it is tempting to find the potential optimum with only four experiments. The result is a matter of chance, however, since a systematic approach for the sequence of the factors to be changed is lacking. Furthermore, not all of the effects that the factors have on the response are determined. The example given here does not investigate the effect of reducing the octane rating at high speed. Nor is there any information on possible interactions: increasing the tyre pressure from 2.2 bar to 2.4 bar changes the fuel consumption by 0.2 L/100 km (0.8 mpg) at a speed of 80 km/h. The small number of experiments does not reveal whether this reduction would be the same, larger or smaller at a speed of 100 km/h; the possible interaction between speed and tyre pressure is not recorded. It is not possible to state with certainty that the minimal fuel consumption of 8.2 L/100 km (34.4 mpg) found in this example really represents the optimum for this process.

From the polemical tone of the language used in this chapter you realise already that the experimental methods presented here are suboptimal. Unfortunately, they are still widely used in practice by scientists and engineers – they are simply entrenched in their minds. The mathematicians/statisticians have not yet managed to convince the experiment planners of the potential of the Design of Experiments method. The author hopes that the following chapters are read by the relevant groups of people and that his "missionary work" falls on fertile ground.

2 The 2^2 Factorial Design to Derive the Methodology and Define the Objectives

In the introduction to the topic of design of experiments in the previous chapter, the term "Factorial Design" method was used. The term *Factorial Design* describes the method best and is therefore frequently used in English-speaking countries. A synonymous term in the international context is: *Design of Experiments (DoE).*

You will see that the systematic design of the experiments is the beginning. After they have been carried out, you obtain results which are evaluated and assessed with statistical methods. And finally, you can then adjust the input variables using the so-called predictive function (the mathematical model of the system) so that the responses assume desired values. A classical example used in teaching here, especially at universities in the USA, is the wooden model of a catapult. The response here is how far a ball (golf ball, softball etc) is thrown. The factors are the adjustment angle at the device, the tensioning of the rubber or the spring, and properties of the ball to be catapulted. With the aid of the predictive function, the adjustment parameters of the catapult can be calculated for a predefined throwing distance. DoE therefore does not necessarily involve achieving a minimum or maximum of a response. Rather, the objective is often to run a system or process at a specific operating point. It is often the case that economic boundary conditions then limit the individual factor settings.

The Factorial Design method claims to ensure a high degree of efficiency in the development and optimisation of products or processes when the approach is systematic and statistical methods are used. From the results of a relatively small number of runs, mathematical models can be obtained which describe the relationship between the input variables (also called influencing factors or the factors) and the responses. When creating a model, it is also important to recognise apparent effects caused by variances in the measured values and to deal with them appropriately. This is done by applying an Analysis of Variance.

Unlike the laws of nature, these models (functional equations) are valid only over a specific range, the so-called experimental space, which is defined by the maximal and minimal values of the factors.

A crucial difference between the Factorial Design method and the OFAT method described in the previous chapter is that several factors are changed at the same time.

Since runs require resources such as staff, time, equipment etc., those responsible for the experiments are torn between two things: on the one hand they have to provide reliable results, for which sufficient data are required. On the other, the effort required for this must be justified. One special attraction of the Factorial Design method is the fact that a minimum number of runs are needed to create a mathematical model which describes the effects and interactions of the factors on the response. Figure 4 shows this in the black box representation.

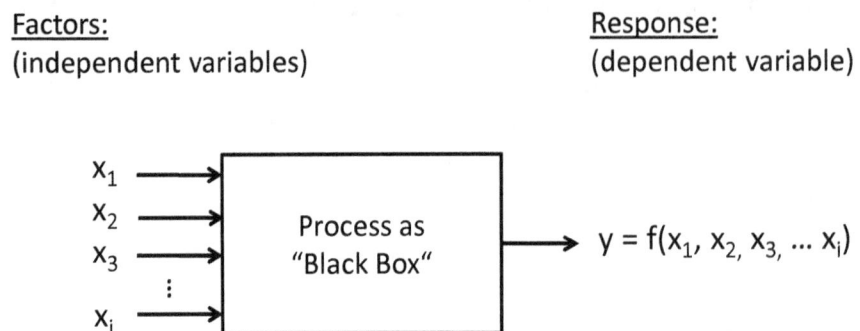

Figure 4: The predictive function: the mathematical model of the Factorial Design method

The factors are not set continuously in small steps, but always in stages. These are also called levels. It may initially sound strange or appear to be trivial that good factorial design results can be achieved with only two levels per factor. But in the course of the following you will see the gain in knowledge which can be achieved even from relatively small amounts of data when a systematic approach is adopted.

Using the example from the previous chapter, which has the product quantity (yield) of a chemical process as the response, we now want to show how an experimental design is created. As has been mentioned, the two factors temperature T and pressure p are to be adjusted in stages. Two stages are defined for each of the two factors: a low level and a high level. For the example stated, these levels are:

$T_{low\ level}$ = 110 °C and $T_{high\ level}$ =150 °C and

$p_{low\ level}$ = 1.4 bar and $p_{high\ level}$ = 3.0 bar.

With this information, a so-called 2^2 factorial design can now be drawn up: two input variables are each set to two levels. The experimental design then consists of four runs in accordance with Table 3. We will later discuss the fact that the runs should be conducted in a random sequence and not according to the standard sequence of the design. This "randomisation" allows noise factors to compensate each other (see Chapter 6).

Trial number	Level factor A: T [°C]		Level factor B: p [bar]		Response M [kg/h]
1	110	(low)	1.4	(low)	M_1
2	150	(high)	1.4	(low))	M_2
3	110	(low)	3.0	(high)	M_3
4	150	(high)	3.0	(high)	M_4

Table 3: Two levels for each factor limit the experimental space of the 2^2 design. Response in this example is the product quantity (yield) in kg/h

The levels of the two factors limit the so-called experimental space within which the factorial design is valid.

The four planned runs are undertaken with the appropriate factor levels and the values of the responses M_1 to M_4 are determined. Each run provides one value of the response as the result, which allows conclusions to be drawn on the effects of the factors involved. Statistical methods are used to investigate the significance of the effects. This involves examining which of the effects observed are statistically "plausible" (= significant) according to the data available. In the next step, we then obtain a functional equation which can be used to predict the response for any combination of factors (intermediate values between the levels) by computation. This naturally applies with the restriction that the model formed from very few data points can reflect reality only approximately, however.

The Factorial Design method does not claim to describe the responses for the whole conceivable range of values of the input variables in accordance with a physical equation. In contrast, the validity is largely limited to the experimental space defined by the specified levels of the factors.

The nomenclature of the designs considered in this book is as follows.

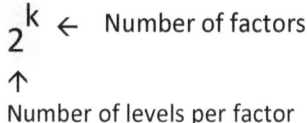

$$2^k \leftarrow \text{Number of factors}$$
$\uparrow$
Number of levels per factor

The 2^2 design shown consists of 4 runs, for example: *k=2* factors each at 2 levels.
Correspondingly, for a 2^3 design with 8 runs: *k=3* factors each at 2 levels.

2.1 Quantitative and qualitative factors

As far as the factors of a design are concerned, a fundamental distinction must be made between quantitative and qualitative factors. The examples so far have involved only quantitative factors. These are (measurable) physical quantities, which can be displayed on an ordinal scale with corresponding unit of measurement, for example temperature in °C, voltages and currents in volts or amperes, or the travel of valves in mm.

In practical factorial design, so-called qualitative factors are also a very frequent occurrence, but their values cannot be expressed as numerical values of an interval scale. They are instead defined by describing characteristics or states with the aid of a nominal scale. Examples are listed in Table 4.

Factor	Stage/level	Stage/level
Catalyst manufacturer	Company A	Company B
Raw material component	Supplier 1	Supplier 2
Reactor condition	Vessel cleaned	Vessel not cleaned
Type of pump	3212	3215
Material	100Cr6	16MnCr5
Test subject or interviewer	Male	Female
Colour	Red	Green

Table 4: Examples of qualitative factors with 2 levels

The last two examples in Table 4 show factors which occur very frequently in factorial designs in medicine/psychology or advertising/marketing. It has been ascertained, for example, that the results of surveys on certain psychological topics can depend significantly on the gender of the interviewer and/or the interviewee. The colour schemes in advertising leaflets can also be crucial factors which can have an effect on responses such as capturing attention.

For reasons of clarity, this book limits itself to experimental designs with two levels per factor. It is obvious that two levels are possibly not sufficient especially for qualitative factors. This is the case when more than two colours or the four points of the compass are to form the factor levels, for example. The IT tools meanwhile available fulfil these requirements so that, with an understanding of the methodology shown here, "higher" designs can also be carried out, where necessary.

As part of the step-by-step derivation of the mathematics for the Factorial Design method in the following chapters, you will see how the levels of the qualitative factors described above in words can be transformed into numerical values.

2.2 Effects of the factors on the response

One intermediate goal forming part of a factorial design is to determine the effects of the factors on the response with the aid of the observations, and to assess their effect on the response with statistical methods.

Example: Product quantity (yield) of a detergent raw material

For a chemical reactor producing a detergent raw material, it is known that the product quantity (yield) obtained per hour M depends on the reactor temperature T (factor A) and the retention time t (factor B) of a particular component in the reactor and can thus be influenced. As part of an experimental design, the effects of the factors stated on the response M are to be investigated.

The levels of the two factors are to be set in accordance with Table 5.

Factor	Low level	High level	Measurement unit
A (Temperature T)	$T_1=130$	$T_2=140$	°C
B (Retention time t)	$t_1=3$	$t_2=4$	h

Table 5: Levels of the factors for the "Product yield of a detergent raw material" experimental design

Table 6 shows the product yields as results of the four runs of a 2^2 experimental design.

| Run number | Levels | | Response M | |
	Factor A: T [°C]	Factor B: t [h]	Yield: [kg/h]	
1	130	3	70	$M_{T1,\,t1}$
2	140	3	72	$M_{T2,\,t1}$
3	130	4	80	$M_{T1,\,t2}$
4	140	4	82	$M_{T2,\,t2}$

Table 6: Measured values of the response for the "Product yield of a detergent raw material" experimental design

In run no. 3, the retention time was increased from 3 to 4 hours at a low temperature level of 130 °C compared to run no. 1. This results in an increase in the product yield M from

$M_{T1,\,t1} = 70\ kg/h$ to $M_{T1,\,t2} = 80\ kg/h$.

This increase in product yield as a result of the increased retention time is called the effect of factor B at a low level of factor A (see Figure 5). The calculation of a factorial effect is defined as the difference of the corresponding values of the response. In the case under consideration it is therefore:

Effect of B at low level of A: $M_{T1, t2} - M_{T1, t1} = 80 - 70 = 10$

(The unit of measurement kg/h is omitted in the following for reasons of clarity).

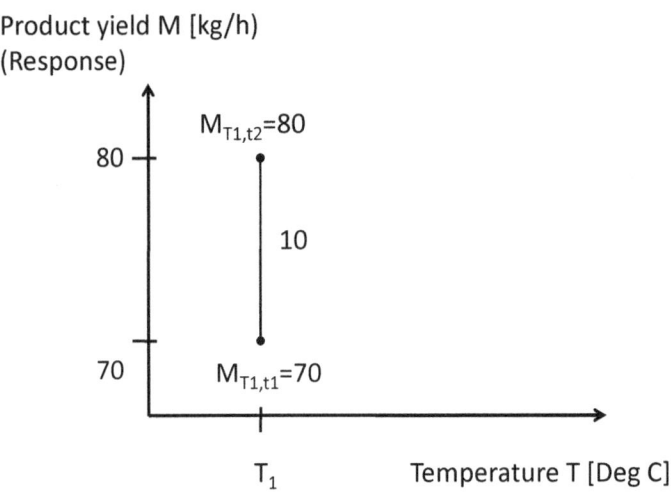

Figure 5: The effect of the retention time t on the yield M
at low temperature level T_1

Next, the effect of factor B (retention time t) at high level of factor A (temperature T) is to be considered. The responses of runs no. 2 and 4 must be compared for this purpose (Figure 6). With $M_{T2, t1} = 72$ and $M_{T2, t2} = 82$, we obtain the desired

Effect of B at high level of A: $M_{T2, t2} - M_{T2, t1} = 82 - 72 = 10$

The result shows that the effect of the retention time on the product yield is independent of the temperature level. In other words: increasing the retention time has the same effect at low temperature as at higher temperature. The yield increases by 10 kg/h in both cases. This is represented in Figure 6 by the parallel connections of the corresponding measuring points (dashed line).

The determination of the effect of temperature (factor A) is carried out in analogy with the determination of the effect of the retention time (factor B):

Effect of A at low level of B: $M_{T2, t1} - M_{T1, t1} = 72 - 70 = 2$

Effect of A at high level of B: $M_{T2, t2} - M_{T1, t2} = 82 - 80 = 2$

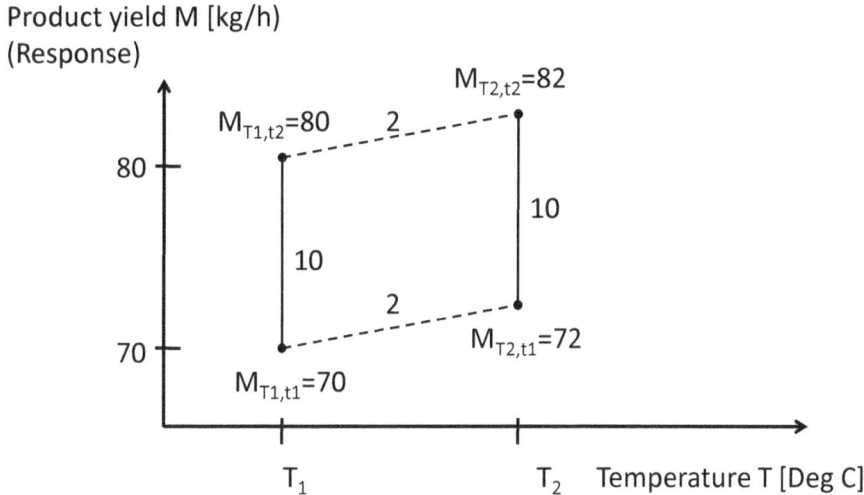

Figure 6: Main effects plot: in this example, the effects of the factors are independent of the level of the other factor

The two effects of the 2^2 factorial design considered in this chapter are called main effects. The case under consideration here with its fictitious data is somewhat atypical, however, because the factors affect the response regardless of the level of the other factor. The main effects plot in Figure 6 has the form of a parallelogram in this case.

In addition to the main effects, there can also be interactions between the factors. What this means and how they are calculated will be shown in the next chapter.

2.2.1 Interactions

The numbers in the example given in Figure 6 were a little contrived, because both factors on both levels of the other factor showed the same effect. Increasing the retention time from the low to the high level led to an increase of 10 kg/h in the product yield on both temperature levels. The temperature increase led to an increase of 2 kg/h in the product yield on both levels of the retention time.

This "harmony" of effects does not always exist in reality. The example will be modified now such that on run no. 4 a product yield of only 76 kg/h (instead of 82 kg/h) would have resulted if both factors had been set to a high level. With the aid of the so-called main effects plot (Figure 7) it can be seen that the increase in the retention time at low temperature level has a greater effect (10 kg/h) than at high level (4 kg/h). This type of behaviour is called interaction.

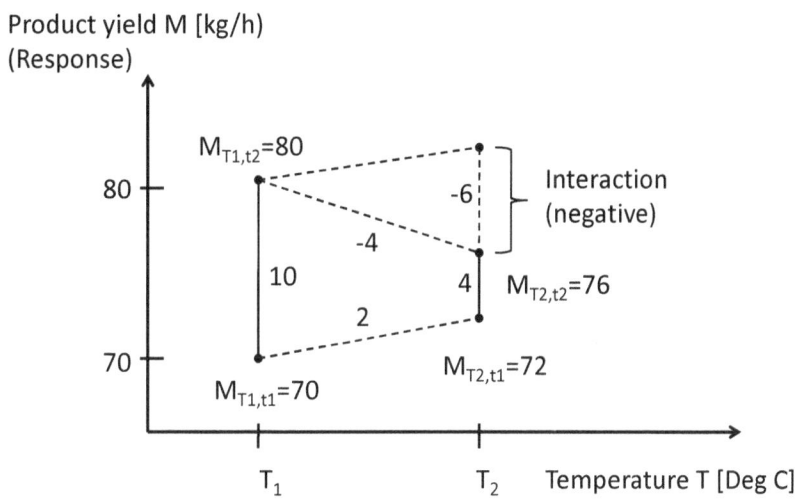

*Figure 7: Main effects plot with negative interaction between
the factors A (temperature T) and B (retention time t)*

An interaction is defined as the difference in the effects of the high and low level of a factor. In the case under consideration, it is called an *AB interaction*[1].

Accordingly, the *AB* interaction is the same as the effect of factor A at high level of factor B minus the effect of A at low level of B. For the example considered, the following applies:

$$AB = (M_{T2,t2} - M_{T1,t2}) - (M_{T2,t1} - M_{T1,t1})$$

---

[1] To distinguish them from the names of the factors, the effects and interactions are written in *italics* here.

Inserting the numerical values results in the *AB* interaction between the factors A and B:

$$(76 - 80) - (72 - 70) = -4 - 2 = -6$$

In the case considered, the *AB* interaction has a negative sign. This means that the longer retention time causes a smaller increase in the product yield at higher temperature.

Is there also a *BA* interaction? That would then be the effect of B at high level of A minus the effect of B at low level of A.

$$BA = (M_{T2,t2} - M_{T2,t1}) - (M_{T1,t2} - M_{T1,t1})$$

You can see that the two relationships *AB* and *BA* are the same. The *AB* interaction is therefore equal to the *BA* interaction. There is therefore only one interaction between the factors A and B, which is called *AB* below.

Interactions can be recognised graphically by the fact that the (broken) straight lines between the experimental points do not run in parallel. Interactions can be positive or negative. In the main effects plots, this can be seen from the slopes of the straight lines. Figure 8 shows the case of a positive interaction. At higher temperature, the retention time here has a greater effect on the increase in the product yield than at low temperature.

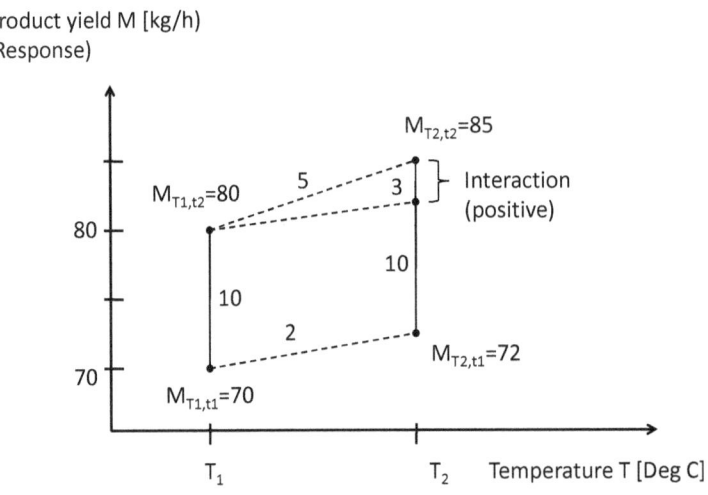

Figure 8: Main effects plot with <u>positive</u> interaction between the factors A (temperature T) and B (retention time t)

With the values given in Figure 8, the *AB* interaction is given by:

$$AB = (M_{T2,t2} - M_{T1,t2}) - (M_{T2,t1} - M_{T1,t1}) = (85 - 80) - (72 - 70) = 5 - 2 = 3$$

2.2.2 Methodology and nomenclature for calculating the effects and interactions

In the earlier chapters, the calculation of the effects and interactions was explained with the aid of an example with numerical values.

A standardised methodology with nomenclature for the designation of the observations, the effects and the interactions is presented below. It can be used to draw up any 2^2 factorial design. In the following chapters, you will see that this methodology can also be applied to "higher" factorial designs.

The four runs of the 2^2 factorial design will be given standardised designations in accordance with Figure 9. First there is the so-called basic test. All factors are at low level here. It is given the designation (g). The designation (1) or (0) for the basic test can also be found in literature. This often causes confusion with the values 1 or 0 when calculating with numbers. In order to prevent this, the basic test is designated as (g) below. For the other three runs, factor names given (in lower case) mean that the corresponding factors must be set to a high level. Factors at low level are not given a designation.

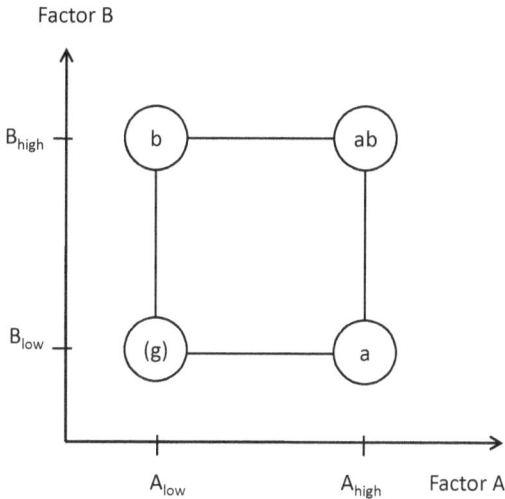

Figure 9: The designations of the four runs of a 2^2 factorial design

With the familiar computational rules to determine the effects and the interaction, the nomenclature stated gives you the following relationships:

Effect of A:

At low level of B:		$a - (g)$
At high level of B:		$ab - b$

Mean effect of A: $\quad\quad\quad\quad\quad \frac{1}{2}(a - (g)) + \frac{1}{2}(ab - b) = \frac{1}{2}(a + ab) - \frac{1}{2}((g) + b)$

Effect of B:

At low level of A:		$b - (g)$
At high level of A:		$ab - a$

Mean effect of B: $\quad\quad\quad\quad\quad \frac{1}{2}(b - (g)) + \frac{1}{2}(ab - a) = \frac{1}{2}(b + ab) - \frac{1}{2}((g) + a)$

AB interaction:

Effect of A at high level of B:		$ab - b$
Effect of A at low level of B:		$a - (g)$

Mean value of the difference of the effects: $\quad \frac{1}{2}(ab - b) - \frac{1}{2}(a - (g)) = \frac{1}{2}((g) + ab) - \frac{1}{2}(a + b)$

It should be noted here that the mean values of the effects between low and high level must be taken as effects and interaction in each case[2].

The following therefore applies:

Effect of a factor	=	Average of all observations where the factor was at <u>high</u> level	-	Average of all observations where the factor was at <u>low</u> level

and

Interaction	=	Average of the difference in the effect of one factor at high and low level of the other factor

With the aid of the equations derived above to calculate the two effects A, B and the interaction AB, the system which determines the sign with which the particular observations enter into the calculation of the effects becomes clear. Using the sign schematic (Table 7), it is simple to compile the equations. When considering the signs to calculate the AB interaction in more detail, you can detect a further useful methodology: the signs to calculate the interaction can be calculated by multiplying the signs of the associated effects.

---

[2] In the introductory numerical examples at the beginning of the chapter, the averaging was omitted for reasons of clarity

All signs in Column I are positive. The sum and the arithmetic mean of the four responses, which will be required later, can thus be calculated. This column is also called the identity column.

Trial		I	Signs of effects		
Number	Name		A	B	AB
1	(g)	1	-1	-1	1
2	a	1	1	-1	-1
3	b	1	-1	1	-1
4	ab	1	1	1	1

$$AB = \frac{1}{2}\left((g) - a - b + ab\right)$$

$$B = \frac{1}{2}\left(-(g) - a + b + ab\right)$$

$$A = \frac{1}{2}\left(-(g) + a - b + ab\right)$$

Table 7: Sign schematic to calculate the effects and the interaction
for the 2^2 factorial design (runs in standard sequence)

The signs of the effects are characterised by -1 and 1 in Table 7. This representation is very suitable to illustrate the arithmetic for calculating the effects. It is also useful to draw up the equations when representing designs with the aid of spreadsheet programs. For reasons of clarity, the internationally customary representation with plus and minus sign for designs with more than two factors will be used in the rest of the book.

The previous numerical example on "product yield" will now show again the calculation of the effects with slightly modified values. Table 8 presents the results of the 2^2 factorial design with the response product yield per hour. Factor A is the reactor temperature T and factor B the retention time t in the reactor.

Run	Yield	Levels	
	[kg/h]	Factor A: T [°C]	Factor B: t [h]
(g)	70	130	3
a	74	140	3
b	80	130	4
ab	82	140	4

Table 8: A further 2^2 factorial design "Product yield"

The effects and the interaction are calculated as follows:

Effect A: $\frac{1}{2}(-(g) + a - b + ab) = \frac{1}{2}(-70 + 74 - 80 + 82) = 3$

Effect B: $\frac{1}{2}(-(g) - a + b + ab) = \frac{1}{2}(-70 - 74 + 80 + 82) = 9$

AB interaction: $\frac{1}{2}(+(g) - a - b + ab) = \frac{1}{2}(+70 - 74 - 80 + 82) = -1$

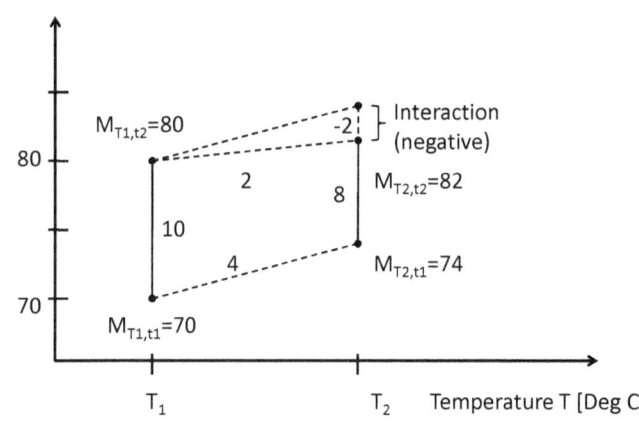

Figure 10: The effects and the interaction of the experimental design from Table 8.
Note: The numerical values for the effects must still be averaged.

Attention is drawn to a little sloppiness in the definitions here. Do not allow yourself to be irritated by the multiple meaning of the following symbols:

    (g), a, b, ab: Run name, observation (response), factor combination
    A, B, AB: Factor designation, effect/interaction

To distinguish between the factor designations and their effects, the latter are printed in *italics* in this book.

Experience has shown that interactions between the parameters occur far more frequently in chemical systems than in mechanical ones. In many cases, this applies to the temperatures and concentrations of the substances involved, which have a reciprocal effect.

The factorial design shown is called a design of 1st order: each factor is set to two levels and the observations are used to determine the coefficients for a linear predictive function.

2.2.3 Graphical representation of the effects and the interaction

It is possible to draw up a very illustrative representation of the effects calculated. For each main effect this is a straight line which is drawn into the so-called main effects plot (Figure 11). The starting and end points of the straight line are formed by the mean values which result from the observations with the relevant factors at low or high level. Table 9 shows the calculation of the starting and end points.

Factor level	Averages of responses	
	A: Temperature	B: Retention time
Low	$\frac{1}{2}((g) + b) = \frac{1}{2}(70 + 80) = 75$	$\frac{1}{2}((g) + a) = \frac{1}{2}(70 + 74) = 72$
High	$\frac{1}{2}(a + ab) = \frac{1}{2}(74 + 82) = 78$	$\frac{1}{2}(b + ab) = \frac{1}{2}(80 + 82) = 81$

Table 9: Averages of the observations (factors at low and high level)

Figure 11 shows the graphical representation of the main effects. Both factors have the effect of increasing the response. From now on the response will be designated with y in preparation for the derivation of the predictive function.

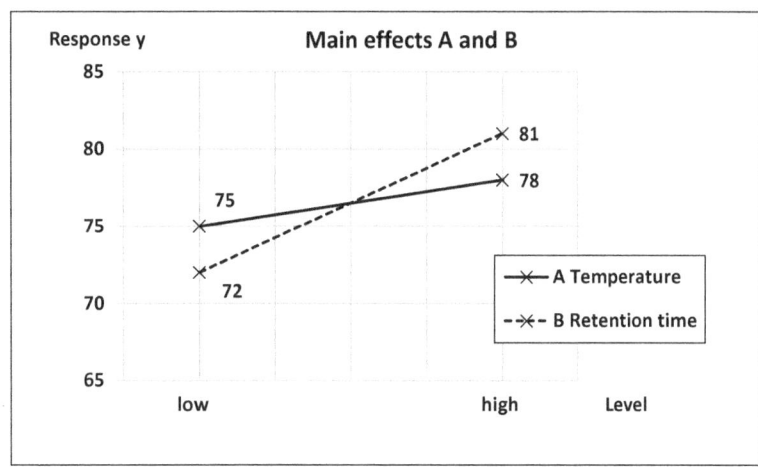

Figure 11: Graphical representation of the main effects A and B in the main effects plot

The graphical representation of the *AB* interaction also consists of two straight lines (Figure 12). The y values which have been the results of the runs at low and high level of A form one straight line (runs *(g)* and *a*). The other straight line is based on the runs *b* and *ab*. In accordance with Table 10, the response y is thus obtained as a function of A with B as the parameter. This representation of the interaction between A and B is called the interaction plot.

It answers the question: What effect has factor A (here the temperature) as a function of parameter B (here the retention time)? You can see from the slopes of the two straight lines that the temperature has a slightly greater effect at shorter retention time.

	A	B	y
B low	-1	-1	$y_{(g)} = 70$
	1	-1	$y_{(a)} = 74$
B high	-1	1	$y_{(b)} = 80$
	1	1	$y_{(ab)} = 82$

Table 10: The observations y define the two straight interaction lines for parameter B

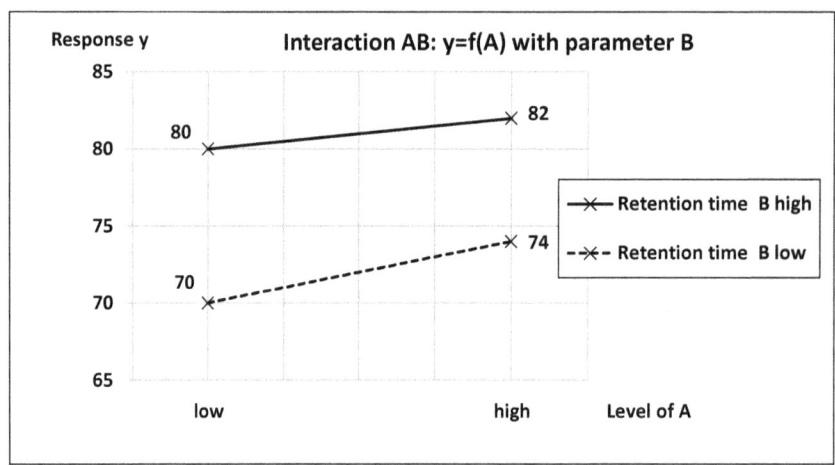

Figure 12: Graphical representation of the AB interaction in the interaction plot;
y=f(A) with parameter B

Conversely, the interaction can also be presented as function $y= f(B)$ with A as the parameter. By comparing Table 10 and Table 11 it is easy to see that the different representations have been generated simply by exchanging the values $y_{(a)}$ and $y_{(b)}$.

	A	B	y
A low	-1	-1	$y_{(g)} = 70$
	-1	1	$y_{(b)} = 80$
A high	1	-1	$y_{(a)} = 74$
	1	1	$y_{(ab)} = 82$

Table 11: The observations y define the two straight interaction lines for parameter A

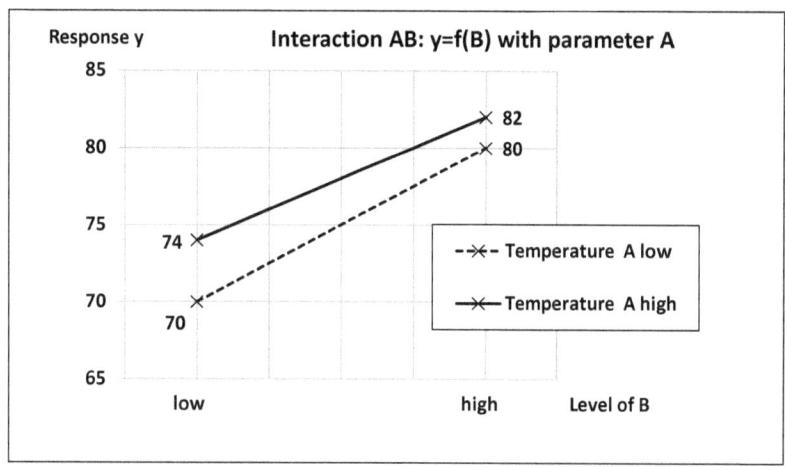

Figure 13: Interaction plot y=f(B) with parameter A

Here as well, you can see from the slope of the two straight lines that the temperature has a slightly greater effect on the response at shorter retention time.

Please note that - as has already been shown - there can only be *one AB* interaction for a 2^2 factorial design. The diagrams above merely show the situation in different representations as a function of parameter B or A.

So what practical use do these interaction plots have? Fundamentally, they serve the qualitative assessment of a suspected interaction. It is easy to see whether it is a positive or negative interaction. And you quickly obtain rough information on the "strength" of the interaction as well.

In the following chapter, the interactions are categorised and a couple of aids are provided for reading the interaction plots. In Chapter 2.2.4 you will use analysis of variance to find out whether observed interactions are statistically significant.

2.2.3.1　Ordinal, disordinal and semi-disordinal interactions

A distinction is made between three types of interaction: ordinal, disordinal and semi-disordinal interactions. This categorisation is described in the following with the aid of the 2^2 factorial design.

Ordinal interactions

An interaction is called ordinal when its absolute value is less than both main effects. In other words: increasing the level leads to an increase in the response or vice versa for both factors. The slopes of the straight lines of the interaction plot have the same sign for *both* parameters. They do not intersect.

This will be explained with the aid of a simple numerical example. For a 2^2 factorial design, four responses *y* as per Table 12 were measured.

Run	Response y
(g)	4
a	6
b	5
ab	9

Table 12: A 2^2 factorial design to demonstrate an ordinal interaction

The straight lines of the interaction plots (Figure 14) do not intersect. This applies to both representations.

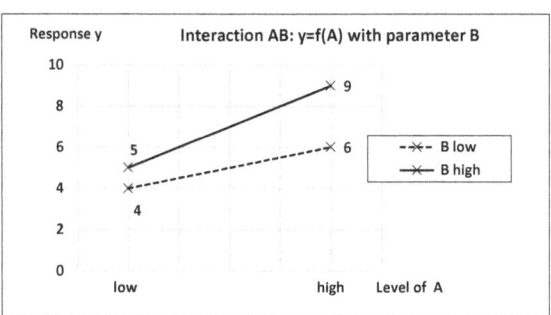 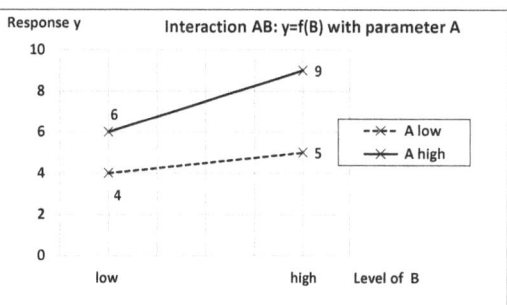

Figure 14: Ordinal interaction: the straight lines have the same sign for both parameters

35

Disordinal interactions

With disordinal interactions, the absolute value of the interaction is greater than the absolute values of both main effects. In the interaction plots, the straight lines intersect in both representations (example in Table 13 and Figure 15).

Run	Response y
(g)	4
a	7
b	8
ab	2

Table 13: A 2^2 factorial design to demonstrate a disordinal interaction

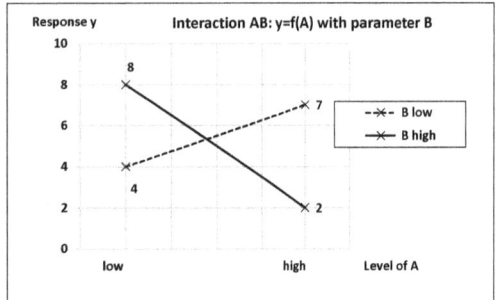 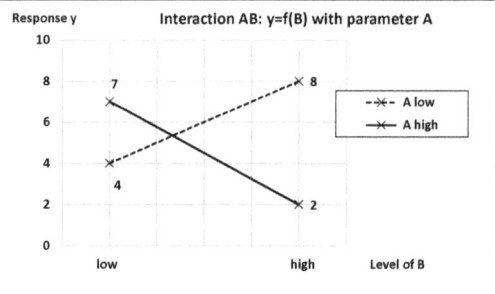

Figure 15: Disordinal interaction: the straight lines intersect for both parameters

Semi-disordinal interactions

For this category, the interaction is greater than one and less than the other main effect (absolute values to be compared). This means there is one disordinal and one ordinal factor. This is the case in the numerical example corresponding to Table 14.

Run	Response y
(g)	2
a	5
b	9
ab	7

Table 14: A 2^2 factorial design to demonstrate a semi-disordinal interaction

In the graphical representation, the straight lines of the interaction plots do not intersect for the ordinal factor. The situation is different for the disordinal factor.

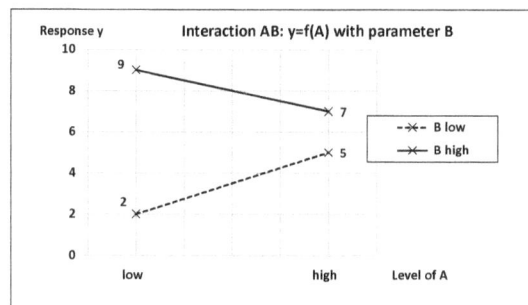 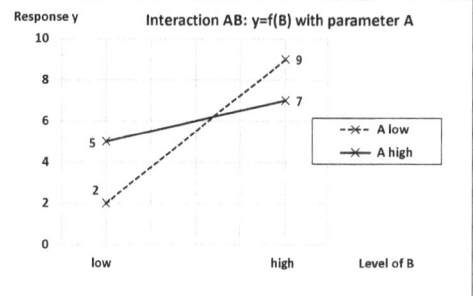

Figure 16: Semi-disordinal interaction: the straight lines intersect for only one of the two parameters

Figure 16 shows that the factor B is ordinal to factor A in the case where *y=f(A)*. The situation is different in the case where *y=f(B)*: A is disordinal to B here.

As the examples in this chapter show, both representations (*y=f(A)* and *y=f(B)*) must be considered in order to categorise the interaction with the aid of the diagrams.

The flow chart in Figure 17 helps to categorise the effects using the slopes of the straight lines.

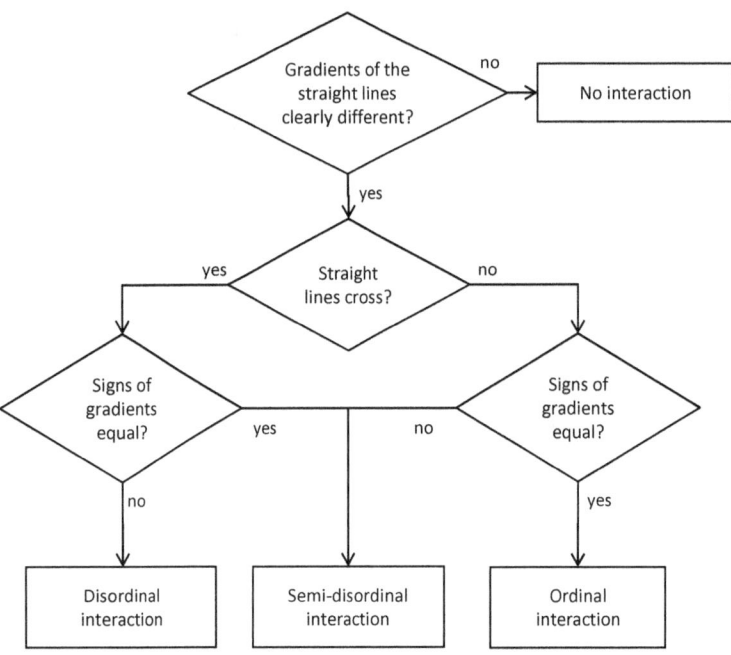

Figure 17: Distinguishing between the three categories of effects using the slopes of the straight lines[3]

In the next chapter, you will learn how a statistical method can be used to reach an objective decision as to whether effects and interactions really do exist or whether the effects observed have to be interpreted differently.

---

[3] According to Jacobs, Bernhard: Einführung in die Versuchsplanung

2.2.4 Significance of effects: Analysis of Variance (F-test)

In the considerations so far, the effects and the interaction calculated from the observations have been considered to actually exist. The focus of this chapter is now to discern whether they really do exist or whether they can be explained by disturbances or noise.

Simply comparing the magnitudes of the effects and the interaction would not be statistically sound, of course. Instead, the Analysis of Variance (ANOVA= *Analysis of Variance*) method is used to assess the significance[4].

The principle of ANOVA and its application to DoE will be briefly explained below and then considered in detail with the aid of numerical examples.

As a reminder: the mean square deviation (or error) between values and their average is a measure for the scatter (=variance) of the values. SS_I designates the so-called sum of squares from the group averages and the overall average. There are two groups here (see below). The overall average is the arithmetic average of all values included in the calculation of the sum of squares. SS_I is thus a measure for the impact of the parameter investigated, so here it is the factor investigated.
SS_R stands for the variance of the individual, measured values about the particular group average and is thus a measure for the proportion of the overall variance which is attributed to the experimental error.

The objective of the Analysis of Variance is to decide with the aid of an F-test whether the factor investigated in each case has a significant effect. The test is carried out at a pre-determined significance level α. The test value F here is formed as shown below:

$$F = \frac{\frac{SS_I}{f_I}}{\frac{SS_R}{f_R}} = \frac{MS_I}{MS_R}$$

f_I and f_R are the degrees of freedom on which the calculation is based. MS_I and MS_R are the averaged sums of squares.

An Analysis of Variance is carried out for each determined effect and each determined interaction of the factorial design. The observations are divided into two groups for this purpose. Group 1 contains the values at which the factor investigated was at a low level. Group 2 comprises the values relating to the high factor level. Table 15 shows the groups for the Analysis of Variance using the example of factor A.

---

[4] See for example Elser, Thomas: Statistik für die Praxis.

	Group values		Group sum	Group mean
Group 1: A low	(g)	b	$(g) + b$	$\dfrac{(g) + b}{2}$
Group 2: A high	a	ab	$a + ab$	$\dfrac{a + ab}{2}$

Table 15: The 2 groups of the Analysis of Variance for factor A of a 2^2 factorial design

The following applies for this Analysis of Variance:

Number of groups: $\qquad I = 2$

Number of degrees of freedom: $\qquad f_I = I - 1 = 1$

Number of values per group: $\qquad J = 2$

The following applies to the sum of squares (between the groups) of factor A:

$$SS_I = SS_A = J \cdot \sum_{i=1}^{I} (\text{group average } i - \text{total average})^2 =$$

$$2 \left(\frac{(g) + b}{2} - \frac{(g) + b + a + ab}{4} \right)^2 + 2 \left(\frac{a + ab}{2} - \frac{(g) + b + a + ab}{4} \right)^2 =$$

$$2 \left(\frac{(g) + b - a - ab}{4} \right)^2 + 2 \left(\frac{a + ab - (g) - b}{4} \right)^2 =$$

$$2 \left(\frac{-2A}{4} \right)^2 + 2 \left(\frac{+2A}{4} \right)^2 = A^2$$

Correspondingly, the following applies for the sum of squares of effect B and the AB interaction:

$$SS_B = B^2$$

$$SS_{AB} = (AB)^2$$

The mean squares are calculated using $MS_I = \frac{SS_I}{f_I}$.

Inserting $f_I = I - 1 = 1$ gives:

$$MS_A = \frac{SS_A}{1} = SS_A$$

$$MS_B = SS_B$$

$$MS_{AB} = SS_{AB}$$

Using the numerical values of the "Product yield" example (Table 8), the mean squares obtained are:

$$MS_A = A^2 = 3^2 = 9$$

$$MS_B = B^2 = 9^2 = 81$$

$$MS_{AB} = (AB)^2 = (-1)^2 = 1$$

The mean square MS_R is now required as a measure for the variance of the values (also called experimental error). In practice, this value is often known from previous experiments or is estimated with the aid of the "higher order" interactions[5]. The present example assumes that the sum SS_R of the mean squares and the degree of freedom f_R on which the calculation is based are known. The values stated are taken to be $SS_R = 1.6$ and $f_R = 4$.

The mean square is thus:

$$MS_R = \frac{SS_R}{f_R} = \frac{1.6}{4} = 0.4$$

The F-test states that the effect investigated is then significant when the test value $F = \frac{MS_I}{MS_R}$ is greater than the critical value $F_{1-\alpha}(f_I; f_R)$ of the F-distribution:

$$F > F_{1-\alpha}(f_I; f_R)$$

For the example under consideration, the F-value of effect A is thus:

$$F_A = \frac{MS_A}{MS_R} = \frac{9}{0.4} = 22.5$$

If the F-test is carried out at a confidence level of 95 %, then the critical value of the F-distribution is[6]:

$$F_{1-\alpha}(f_I; f_R) = F_{0.95}(1; 4) = 7.7$$

The result of the F-test for effect A is: since the calculated F-value ($F_A = 22.5$) is greater than the critical value of the F-distribution ($F_{0.95}(1; 4) = 7.7$), it is assumed that the effect A is significant with a significance level of 5 %.

The F-tests are now similarly carried out for effect B and the interaction AB. The results are shown in Table 16.

---

[5]For factorial designs with more factors (see chapters below), 3-factor and multi-factor interactions exist (in theory). They are rarely significant, however, and can therefore be used to estimate the experimental variance.

[6] Note the complementary quantities confidence level 1-α and significance level α.

Note: It has been assumed in this example that the experimental variance is known. In practice, it is estimated from interactions which are presumed not to be significant (see subsequent chapters). If it can be economically justified, factorial designs are conducted twice, for example. In these cases, there are two runs for every level combination, and their results are used to estimate the experimental error (see Chapter 4.2).

The mean square deviations between the group averages and the overall average are important for the arithmetic of the Analysis of Variance. One degree of freedom is '"lost", as is known from the calculation of the variance of a sample. In this case of the ANOVA with only two groups, this leads to a situation where the mean squares MS_I are equal to the sum squares SS_I, as shown, because the degree of freedom is $f_I = 1$. The mean squares MS_R for the error estimation are calculated from the sum of the mean squares of the values used, divided by their number.

Factor	Effects Product yield [kg/h]	$SS_I=MS_I$	$F=MS_I/MS_R$	p-value	F-test result
A	3.0	9.0	22.50	0.0090	Significant
B	9.0	81.0	202.50	0.0001	Significant
AB	-1.0	1.0	2.50	0.1890	

Table 16: Analysis of Variance (ANOVA) for the data from Table 8: a) Comparison F-values with F-critical value or b) Comparison p-values with $\propto = 0.05$

As a result of this Analysis of Variance, we note that the main effects A and B are considered to be significant, whereas the AB interaction is not. The observed effect of AB is attributed to the experimental error.

p-value

A further note: As has been described, with the F-test, the significance of an effect is confirmed by the fact that the corresponding F-value of a factor is higher than the critical value of the F-distribution. In many publications and software packages on DoE, the test result is determined with the aid of the so-called p-value. The p-value is the area under the probability density function of the F-distribution which is to the right of the corresponding F-value of an effect The rule is: if the p-value of an effect is lower than the significance level $\propto$ set for the test, then the effect is confirmed as being significant. The examples in this book use $\propto = 0.05$, as is frequently the case in practice, i.e. an effect is considered to be significant if $p < 0.05$ (see Table 27).

Some software packages use a t-test instead of the F-test[7]. The procedure is principally the same.

---

[7] More information about the t-test, F-test, ANOVA and degrees of freedom: See also Elser, Thomas: Statistik für die Praxis.

2.3 Evaluation of the factorial design: the predictive function - the mathematical model

The derivation of the expressions to calculate effects and interactions and to assess their significance was explained in detail in the preceding chapters using a 2^2 factorial design. The next step now shows how these results can be used to formulate an equation for the function $y = f(A, B, AB)$. This equation can then be used to calculate the response y for any combination of the factor settings (between the levels). A linear mathematical model is thus available to describe the process investigated or the system considered, in order to then be able to set it to a desired operating point.

The designs considered so far operate with two levels for each factor. For these factorial designs of 1st order, the response y has a linear dependence on each factor.

First, the nature of the interaction will be explained again: For a low level of the factor, the interaction must be subtracted from the effect, for a high level it must be added to it. For negative interactions, this operation is reversed by the sign of the numerical value, of course. The following expressions and diagrams apply in general, however, i.e. for negative and positive interactions.

For calculations in the factorial design, four scenarios are now considered by way of example in order to show how the functional equation is formed so as to be able to calculate with any level combination in the experimental space[8]. The functional equation is sometimes also called the predictive equation. More suitable are the similarly common designations "predictive function" and "prediction model".

1st scenario: Calculation from corner to corner (Figure 18)

The main effects must be inserted in full here and the interaction added or subtracted.

$(g) + B - AB = 70 + 9 - (-1) = 80 = b$ B is increased with A at <u>low</u> level

$a + B + AB = 74 + 9 + (-1) = 82 = ab$ B is increased with A at <u>high</u> level

$(g) + A - AB = 70 + 3 - (-1) = 74 = a$ A is increased with B at <u>low</u> level

$b + A + AB = 80 + 3 + (-1) = 82 = ab$ A is increased with B at <u>high</u> level

---

[8] According to Engelmann, H.-D., Erdmann H.-H., Simmrock, K.H.: Planen und Auswerten von Versuchen

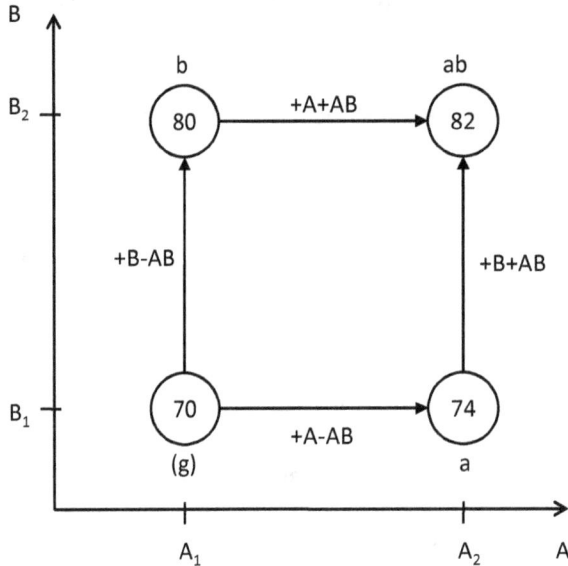

Figure 18: Calculations in the factorial design: from corner to corner

2nd scenario: Calculation from the corner to the centre (Figure 19)

Half of each of the two main effects must be inserted here and the interaction added or subtracted.

Note: If factors on the medium level of the other factor are increased, no interaction shall be inserted (see 2nd step in this example).

1st step: B is increased by 50 % with A at <u>low</u> level:

$$(g) + \frac{B}{2} - \frac{AB}{2} = 70 + \frac{9}{2} - \frac{-1}{2} = 75$$

2nd step: B is increased by 50 % with A at <u>medium</u> level:

$$75 + \frac{A}{2} = 75 + \frac{3}{2} = 76.5 = \bar{y}$$

Alternatively, it is also possible to take the routes marked by the dashed lines in Figure 19:

1st step   A is increased by 50 % with B at low level

$$(g) + \frac{A}{2} - \frac{AB}{2} = 70 + \frac{3}{2} - \frac{-1}{2} = 72$$

2nd step: B is increased by 50 % with A at medium level:

$$72 + \frac{B}{2} = 72 + \frac{9}{2} = 76.5 = \bar{y}$$

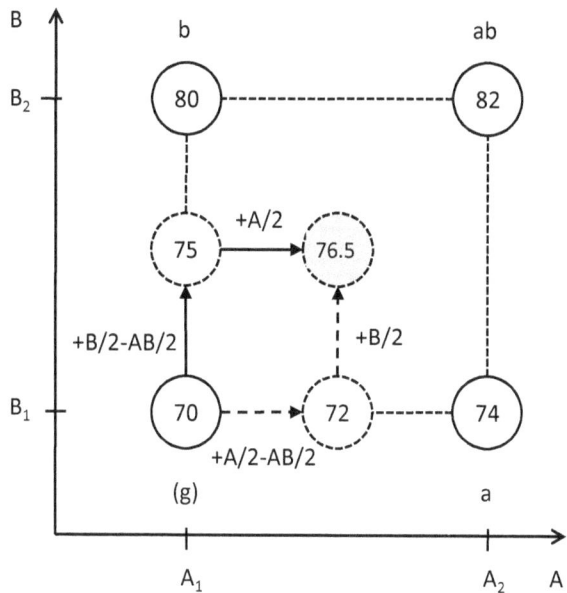

Figure 19: Calculations in the factorial design: from the corner to the centre

You can see that it is possible to arrive in the centre $\bar{y} = 76.5$ of the factorial design via both routes. This centre corresponds to the arithmetic average of the responses from the four runs:

$$\bar{y} = \frac{(g) + a + b + ab}{4} = \frac{70 + 74 + 80 + 82}{4} = 76.5$$

3rd scenario: Calculation from the centre to the corner (Figure 20)

Half of each of the two main effects must again be inserted and the interaction added or subtracted.

1st step: <u>B</u> is increased by 50 % with <u>A</u> at <u>medium</u> level: $\bar{y} + \frac{B}{2} = 76.5 + \frac{9}{2} = 81$

2nd step: <u>A</u> is increased by 50 % with <u>B</u> at <u>high</u> level: $81 + \frac{A}{2} + \frac{AB}{2} = 81 + \frac{3}{2} + \frac{-1}{2} = 82 = ab$

or via the routes represented by dashed lines in Figure 20:

1st step: <u>A</u> is increased by 50 % with <u>B</u> at <u>medium</u> level: $\bar{y} + \frac{A}{2} = 76,5 + \frac{3}{2} = 78$

2nd step: <u>B</u> is increased by 50 % with <u>A</u> at <u>high</u> level: $78 + \frac{B}{2} + \frac{AB}{2} = 78 + \frac{9}{2} + \frac{-1}{2} = 82 = ab$

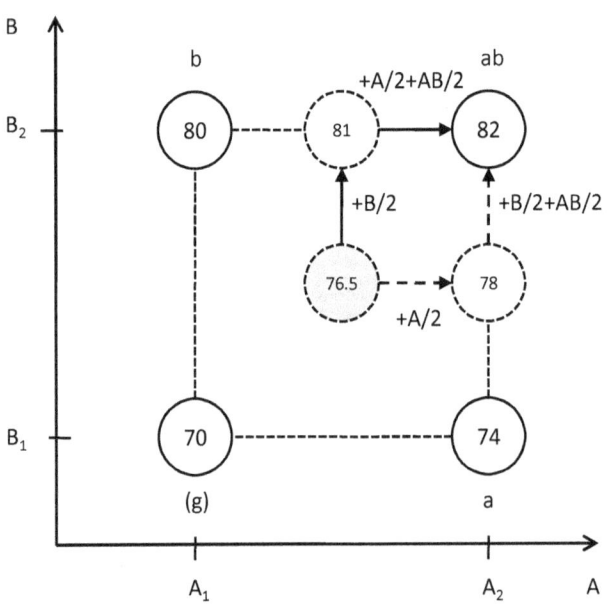

Figure 20: Calculations in the factorial design: from the centre to the corner

4th scenario: Calculation from the centre to arbitrary intermediate values of A and B (Figure 21)

In this case, the objective of being able to calculate with any value of the input variables has already been almost achieved. We now assume that both factors are simultaneously set to levels which are between their mean values and their maximum values.

The values for the pro rata increases of the levels are taken to be $x_A = 0.3$ and $x_B = 0.2$.

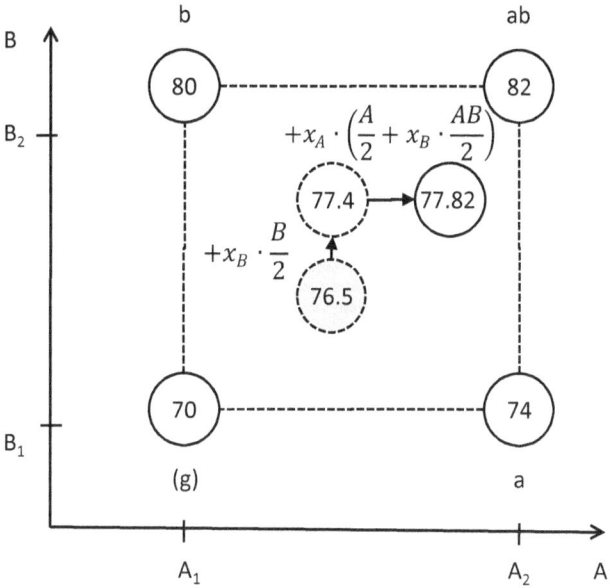

Figure 21: Calculations in the factorial design: from the centre to arbitrary points

From the centre of the factorial design, the calculation is initially towards factor B in accordance with Figure 21, but not to the high level of B, but only on a pro rata basis in accordance with x_B:

$$x_B \cdot \frac{B}{2} = 0.2 \cdot \frac{9}{2} = 0.9$$

$$y = \bar{y} + x_B \cdot \frac{B}{2} = 76.5 + 0.9 = 77.4$$

Starting from there, the effect of factor A on this level is to be taken in accordance with x_B. This means that the AB interaction is only to be taken into account on a pro-rata basis in accordance with x_B:

$$\frac{A}{2} + x_B \cdot \frac{AB}{2}$$

This would be the effect of A if we were to calculate to the end of the experimental space (high level of A).

In reality, however, the calculations here should only go up to the effect of A in accordance with x_A. We therefore have to write:

$$x_A \cdot \left(\frac{A}{2} + x_B \cdot \frac{AB}{2} \right) = 0.3 \cdot \left(\frac{3}{2} + 0.2 \cdot \frac{-1}{2} \right) = 0.42$$

With these intermediate results, it is now possible to calculate from the centre of the design:

$$y = 76.5 + 0.9 + 0.42 = 77.82$$

or using letters:

$$y = \bar{y} + x_B \cdot \frac{B}{2} + x_A \cdot \left(\frac{A}{2} + x_B \cdot \frac{AB}{2} \right)$$

Rearranging slightly, this gives the predictive function of the 2^2 factorial design:

$$y = \bar{y} + \frac{A}{2} x_A + \frac{B}{2} x_B + \frac{AB}{2} x_A x_B$$

2.3.1 Normalised representation of the predictive function

The predictive function can be represented in a slightly more convenient way by introducing the coefficients b_0 to b_3. With $b_0 = \bar{y}$, $b_1 = \frac{A}{2}$, $b_2 = \frac{B}{2}$ and $b_3 = \frac{AB}{2}$, the following function is obtained:

$$y = b_0 + b_1 x_A + b_2 x_B + b_3 x_A x_B$$

Software products for DoE and the tables printed in the following chapters often show the coefficients of the predictive function in addition to the effects.

When undertaking calculations in the factorial design, it has proved to be worthwhile to select the interpolation factors x_A and x_B in such a way that they have values between -1 and 1:

$$1 \le x_A \le 1$$
$$1 \le x_B \le 1$$

To this end, a coordinate system as per Figure 22 is introduced whose origin lies in the centre of the factorial design. Both axes have the same dimensionless scale; the experimental space can therefore be represented by a square.

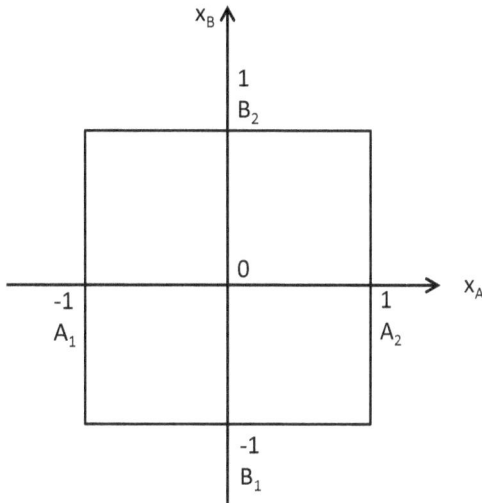

Figure 22: Normalised system of coordinates for the experimental space

The arithmetic for the interpolation factors is as follows:

$$x_A = \frac{A^* - \frac{1}{2}(A_2 + A_1)}{\frac{1}{2}(A_2 - A_1)} \qquad\qquad x_B = \frac{B^* - \frac{1}{2}(B_2 + B_1)}{\frac{1}{2}(B_2 - B_1)}$$

In these equations, A^* and B^* are the levels for which the response is to be calculated. A_1, B_1 and A_2, B_2 are the values of the low or high levels of the two factors.

The predictive function shall now be examined (numerical values corresponding to Table 8). A temperature level of $A^* = 136.5\,°C$ and a retention time of $B^* = 3.6\,h$ shall be set. Which yield y is to be expected?

The result for the normalised quantities is:

$$x_A = \frac{A^* - \frac{1}{2}(A_2 + A_1)}{\frac{1}{2}(A_2 - A_1)} = \frac{136.5 - \frac{1}{2}(140 + 130)}{\frac{1}{2}(140 - 130)} = 0.3$$

and

$$x_B = \frac{B^* - \frac{1}{2}(B_2 + B_1)}{\frac{1}{2}(B_2 - B_1)} = \frac{3.6 - \frac{1}{2}(4 + 3)}{\frac{1}{2}(4 - 3)} = 0.2$$

When inserted into the predictive function, this gives:

$$y = \bar{y} + \frac{A}{2}x_A + \frac{B}{2}x_B + \frac{AB}{2}x_A x_B = 76.5 + \frac{3}{2} \cdot 0.3 + \frac{9}{2} \cdot 0.2 + \frac{-1}{2} \cdot 0.3 \cdot 0.2 = 77.82$$

This result corresponds to the value which was described in the step-by-step derivation of the predictive function as scenario 4 in the last chapter.

A further example will confirm the consistency of the system again. The response y which results when both factors are at a high level is to be calculated. With $x_A = x_B = 1$, the following results in this case:

$$y = \bar{y} + \frac{A}{2} + \frac{B}{2} + \frac{AB}{2} = 76.5 + \frac{3}{2} + \frac{9}{2} + \frac{-1}{2} = 82$$

This value corresponds to the response of the run ab – the calculation model is therefore free from contradiction so far.

Since the Analysis of Variance for the exemplary data (Table 8) showed that, although the effects A and B were significant, the AB interaction was not (Table 16), the predictive function must be adapted accordingly: this is done by setting the effect values of the effects which are not significant to 0.

With $AB = 0$ the term $\frac{AB}{2}x_A x_B$ is zero and the following predictive function is obtained:

$$y = b_0 + b_1 x_A + b_2 x_B$$

With the factor values $A^* = 136.5\ °C$ and $B^* = 3.6\ h$ or $x_A = 0.3$ and $x_B = 0.2$, the following response is obtained[9]:

$$y = \left(76.5 + \frac{3}{2} \cdot 0.3 + \frac{9}{2} \cdot 0.2\right) kg/h = 77.85\ kg/h$$

Representing the observations as the corners of a square (see Figure 21, for example) is slightly misleading. As shown in Figure 23, it becomes clear that the diagram to represent the physical experimental space is fixed by the scale of the (physical) factor level. The so-called normalised experimental space has the same scale for both axes and can therefore be represented as a square.

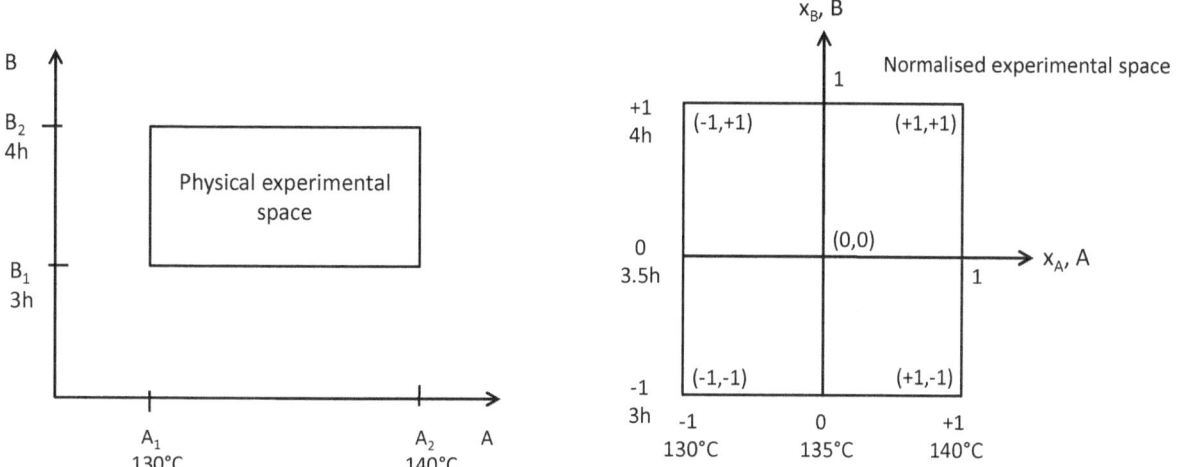

Figure 23: The relationship between physical and normalised experimental space

Owing to the linear approach used for the predictive function for 1st order designs (factors are set to two levels), maxima and minima of the response can be located only at the corners of the experimental space. As has already been mentioned in the introduction, optima of processes can be located – often as a result of economic considerations – at arbitrary locations in the experimental space.

The results of this and the previous chapters are shown in compact form in Chapter 2.5 as a computing scheme for 2^2 full factorial designs. Exemplary realisations of the equations using the Microsoft Excel® and OpenOffice Calc® spreadsheet programs are available for download (see Chapter 9). By entering the four observations y and the associated factor levels it is thus possible to "play" with the 2^2 factorial design. The calculations concern the effects and the coefficients of the predictive function and the F-values for the Analysis of Variance. The effects and the interaction are shown in diagrammatic form. The responses to be expected can be calculated for "arbitrary" settings of the factors.

Alternatively, statistics software with DoE functionality is available, as is very specialised DOE software. Test versions with a time limit are also frequently on offer (see Chapter 10).

---

[9] Note: The term response is always used when the computational result of the predictive function is meant

For those interested in the mathematics:

The predictive function is a linear function with two independent variables x_A and x_B. The halved effects A, B and AB are the coefficients of this function. As shown, the function was developed from the data of the four runs. In mathematical terms, a twofold linear regression of x_A and x_B with regard to the response y was carried out. The predictive function obtained describes a plane, the regression plane. It is now possible to calculate by way of approximation the resulting response on the regression plane for any factor combinations. With only four runs, this approximation can naturally be only very rough.[10]

---

[10] How "well" the predictive function represents reality should be determined with the aid of an analysis of residuals. See also Kleppmann, Wilhelm: Taschenbuch Versuchsplanung.

2.3.2 Hidden effects

The previous chapter used an exemplary 2^2 factorial design to show how the predictive function can be compiled from the observations. Within the experimental space, a linear relationship was assumed between the factors investigated and the response. This does not always have to be the case, however. On the contrary, there are non-linear relationships between factors and responses as well. In this chapter, this problem is illustrated using an example: so-called hidden effects involve the risk of the observations being misinterpreted. How hidden effects are recognised and how to deal with them is explained in the following.

Example: Yield of a chemical reaction

For the synthesis of a chemical product, the impact of temperature A and pressure B on the yield [11] are to be investigated using a 2^2 factorial design.

Table 17 and Table 18 show the levels of the factors and the observations.

Response: yield [%]				Factor levels	
Factor		Measurement unit	Factor type	Low	High
A	Temperature	°C	Quantitative	100	140
B	Pressure	bar	Quantitative	1.5	2.5

Table 17: Levels of the "Yield" 2^2 factorial design

Pressure B		Temperature A			
		100	(-1)	140	(+1)
1.5	(-1)	(g)	79.62	a	77.48
2.5	(+1)	b	74.64	ab	79.64

Table 18: Observations of the "Yield" 2^2 factorial design

To compile the main effects plot, the mean values of the observations are calculated using the equations from Table 9, where the factors were set to a high and a low level in each case. The runs are shown in Table 19.

---

[11] In chemistry, yield is defined as the ratio of the quantity of the product obtained and the maximum (without any losses) amount of substance to be obtained in theory.

	Main effect	
Factor level	A	B
	Temperature	Pressure
Low	77.13	78.55
High	78.56	77.14

Table 19: The main effects of A and B (yields in %)

These results can now be used to compile the diagrams in Figure 24. As has already been shown, the corner points for the interaction plot result directly from the observations.

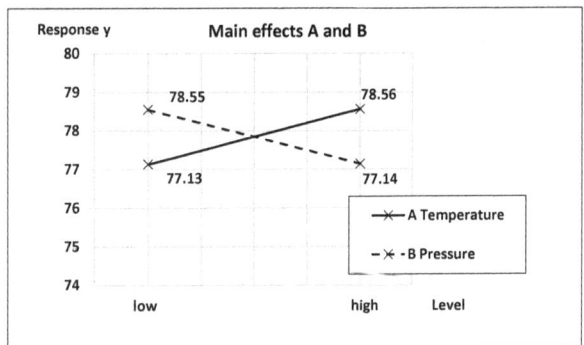 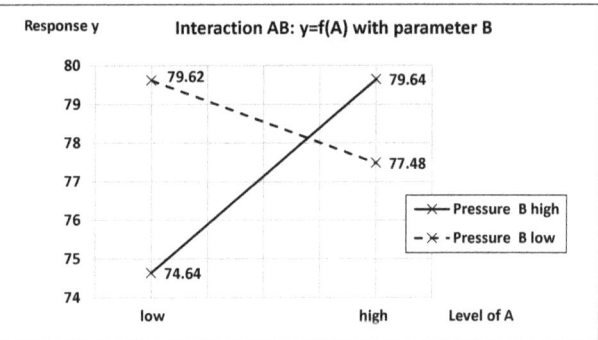

Figure 24: Main effects plot and interaction plot of the "Yield" 2² factorial design

From the graphical representation, it is easy to see that the factors have a mutual influence. This in itself is nothing special and has already been shown with the aid of previous examples. It is interesting here to consider the numerics of the effects and the interaction, however.

The calculation of the effects results in:

Temperature A: $\qquad \frac{1}{2}(-(g) + a - b + ab) = \frac{1}{2}(-79.62 + 77.48 - 74.64 + 79.64) = +1.43\,\%$

Pressure B: $\qquad \frac{1}{2}(-(g) - a + b + ab) = \frac{1}{2}(-79.62 - 77.48 + 74.64 + 79.64) = -1.41\,\%$

The following applies to the interaction between temperature and pressure:

AB interaction: $\qquad \frac{1}{2}(+(g) - a - b + ab) = \frac{1}{2}(+79.62 - 77.48 - 74.64 + 79.64) = +3.75\,\%$

The ordinal AB interaction is conspicuous here. It is much stronger than the main effects A and B. If the interaction were to be taken as an estimate for the experimental variance, the two main effects would not be considered to be significant in the F-test. This would be a risky - and in this case incorrect - interpretation, however.

It is therefore more probable that the non-significance of the main effects was only "spurious". It has so far been assumed that the factors have a linear effect on the response within the experimental space. If the effect were in accordance with Figure 25, however, the issue would be one of so-called hidden effects *A* and *B*. In this fictitious example, the response would result in the maximum yield of 81 % at low pressure and a temperature of approx. 116 °C.

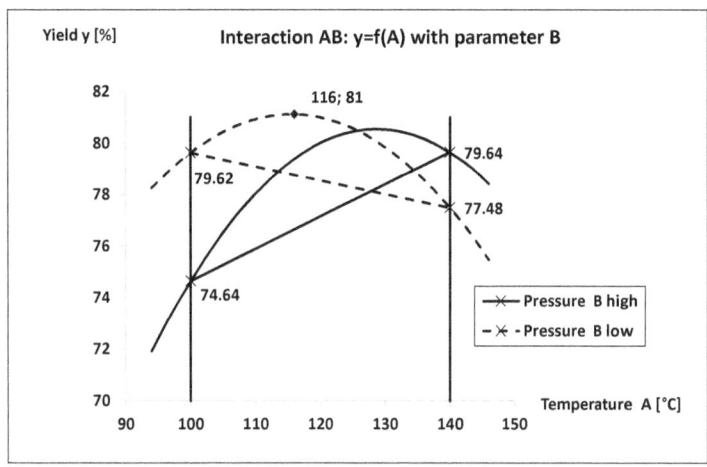

Figure 25: Hidden effects

It is easy to understand that repeat runs with the factor levels low and high would not provide any information on the position of the maximum. Instead, additional runs with other settings must be done. A tried and tested procedure here is to expand the previous factorial design by one or more "center point run"s.

The center point runs, which are usually conducted several times, say a great deal about the extent of the nonlinearity. The difference between the mean value of the center point runs and the mean value of the other runs is calculated. This difference is a measure for the deviation from linearity. It is called *lack of fit*. The significance of the deviation is determined by a statistical test.

Figure 26 shows an extension of the factorial design by four "star point runs". As can be seen, four additional runs were grouped around the centre of the design in the form of a star (a so-called *star design*). These types of designs are called central composite designs. The distance α to the centre of the design is cleverly selected to suit the physical circumstances. The rule of thumb is:

$$\alpha = (Number\ of\ runs\ (g), a, b, ab \dots)^{1/4}$$

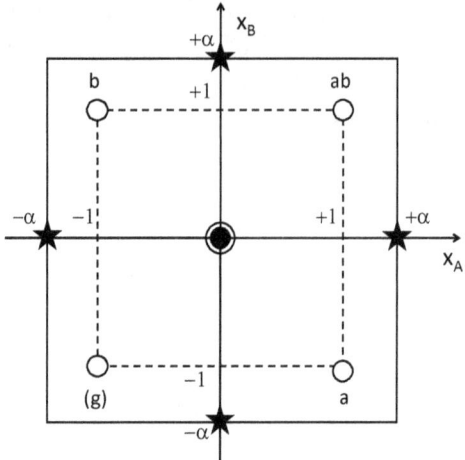

Figure 26: Extension of the factorial design by center point runs and star point runs
(Central Composite Design)

For the present case, the experimental space can be graphically represented by Figure 27. The eight levels to be set are located on a circle at a normalised distance $\sqrt{2}$ around the center point run. This is called a central composite rotatable design.

Rotatable in this context means that the observations produced by uniform "rotation" of the levels of the runs on the circle always have the same information content. This is because the separations of the run levels and the boundaries of the experimental space remain the same.

In the present case, rotatability is ensured with $\alpha = (4)^{1/4} = \sqrt[2]{2} \approx 1.432$.

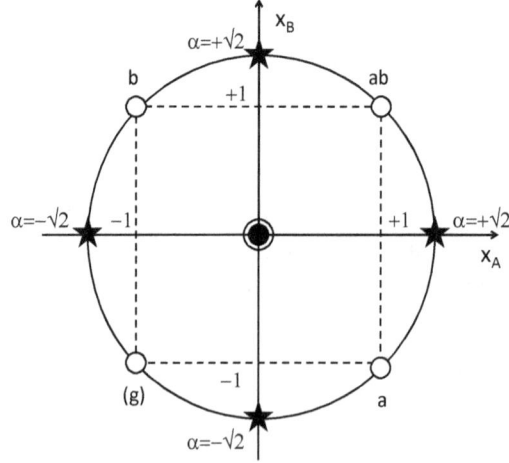

Figure 27: Central composite rotatable design

For physical or technical reasons, it is not always possible to set α according to this rule. If the factor is temperature, for example, a setting above the normalised value 1 (high level) may not be possible because, in this case, an undesired chemical reaction would start or the installation could be damaged at higher temperatures. Star point runs are obviously impossible for qualitative factors.

Conclusion: The determination of α for the star point runs according to the equation is merely an indication for a central composite rotatable design. In reality, individual α are often required for each factor. This may be down to physical/technical circumstances or economic reasons.

The additional star point runs thus expand the experimental space by a factor of $\sqrt{2}$. This means that the levels of the star point runs exceed or remain below the previous factor levels. For the present example, these are 28.3 °C for temperature A and then correspondingly approx. 0.7 bar for pressure B. How accurately the values stated in Table 20 can be set depends on the technical features of the system.

	Factor runs		Center point run(s)	Star point runs	
	Low (−1)	High (+1)		Low (−α)	High (+α)
Temperature A [°C]	100	140	120.0	71.7	168.3
Pressure B [bar]	1.5	2.5	2.0	0.8	3.2

Table 20: The levels for the central composite rotatable design of the "Yield" example

The factorial design, supplemented by star and centre point runs, could then look as presented in Table 21[12].

With the results of the star point runs, it is now possible to determine the predictive function for the response, which contains quadratic terms as well as the linear ones. The full quadratic approach for 2 variables would then have the form (2nd order factorial design; quadratic model):

$$y = b_0 + b_1 x_A + b_2 x_B + b_3 x_A^2 + b_4 x_B^2 + b_5 x_A x_B$$

In conclusion, the rule to recognise hidden effects can be formulated as follows: If an interaction appears to be significant and the associated main effects do not, then one has to expect hidden effects and the above-mentioned measures have to be taken.

---

[12] The runs No. 1 to 4 are sometimes called factor runs to distinguish them from central and star point runs.

Run		Factor level	
Number	Name	A	B
1	(g)	-	-
2	a	+	-
3	b	-	+
4	ab	+	+
5	Center point run	0	0
6	Center point run	0	0
7	Star point run: $-\alpha\ for\ A$	$-\alpha$	0
8	Star point run: $+\alpha\ for\ A$	$+\alpha$	0
9	Star point run: $-\alpha\ for\ B$	0	$-\alpha$
10	Star point run: $+\alpha\ for\ B$	0	$+\alpha$
11	Center point run	0	0
12	Center point run	0	0

Table 21: A central composite rotatable design for a 2^2 factorial design.
For clarity, the levels have been denoted by +/ - instead of 1/-1.

The factorial designs dealt with later in this book and the arithmetic of the predictive functions assume that the responses run linearly between the levels (designs of 1st order). In reality, this is approximately correct in very many cases, because the experimental spaces can be kept as small as possible by clever choice of the levels.

Designs of a higher order inevitably require more computational work. In this case, we draw attention to professional DoE software which processes the data and provides the graphical representations of the results (see Chapter 10).

2.4 2^2 example "Surface roughness of turned parts"

The example calculated below shows the practical procedure step-by-step. The considerations are explained and these then lead to the development of a calculation model in the following chapter. The very simple arithmetic can easily be implemented using a spreadsheet program. [13]

Task: Surface roughness of turned parts

One characteristic determining the quality when manufacturing parts on a lathe is the surface roughness of the fabricated pieces. The assumption is that the cutting depth (in mm) which is set and the feed (in mm per revolution) have a significant effect on the surface roughness. The natural variance of the process is assumed to be known from earlier experiments: $MS_R = 0.105$ (with $f_R = 4$).
The F-test as part of the Analysis of Variance to assess the significance of the effects is to be based on a significance level of 5 %. The task is to use the predictive function to help calculate which surface roughness is to be expected when the machine is set to a cutting depth of 0.8 mm and a feed of 0.04 mm per revolution. [14]

1st step:　　　Plan and conduct experiments

Table 22 and Table 23 show the factorial design and the observations:

	Response: surface roughness [µm]			Factor levels	
	Factor	Measurement unit	Factor type	Low	High
A	Cutting depth	mm	Quantitative	0.5	1
B	Feed	mm	Quantitative	0.005	0.1

Table 22: Levels of the "Surface roughness of turned parts" 2^2 factorial design

Feed B		Cutting depth A			
		0.5	-	1	+
0.005	-	(g)	2.8	a	3.5
0.1	+	b	3	ab	5.6

Table 23: Observations of the "Surface roughness of turned parts" 2^2 factorial design

---

[13] MS Excel®/OpenOffice Calc®-files for download, see Chapter 9.

[14] Following the example of Ament, Ch.: Eine Einführung in die statistische Versuchsplanung.

The effects and the interaction are calculated with the aid of the equations in column 1 of Table 24. The *main effects plot* and the *interaction plot* in accordance with Table 25 and Table 26 serve as illustrative graphical representations.

Effects and interaction	Coefficients of the predictive equation
	$b_0 = \bar{y} = 3.725$
$A = \dfrac{1}{2}(-(g) + a - b + ab) = \dfrac{1}{2}(-2.8 + 3.5 - 3 + 5.6) = 1.65$	$b_1 = \dfrac{A}{2} = 0.825$
$B = \dfrac{1}{2}(-(g) - a + b + ab) = \dfrac{1}{2}(-2.8 - 3.5 + 3 + 5.6) = 1.15$	$b_2 = \dfrac{B}{2} = 0.575$
$AB = \dfrac{1}{2}(+(g) - a - b + ab) = \dfrac{1}{2}(+2.8 - 3.5 - 3 + 5.6) = 0.95$	$b_3 = \dfrac{AB}{2} = 0.475$

Table 24: Calculation of the effects and the interaction of the
"Surface roughness of turned parts" 2^2 factorial design

Factor level	Main effects	
	Cutting depth A	Feed B
-1 (low)	$\frac{1}{2}\left((g)+b\right)=\frac{1}{2}(2.8+3)=2.90$	$\frac{1}{2}\left((g)+a\right)=\frac{1}{2}(2.8+3.5)=3.15$
1 (high)	$\frac{1}{2}(a+ab)=\frac{1}{2}(3.5+5.6)=4.55$	$\frac{1}{2}(b+ab)=\frac{1}{2}(3+5.6)=4.30$

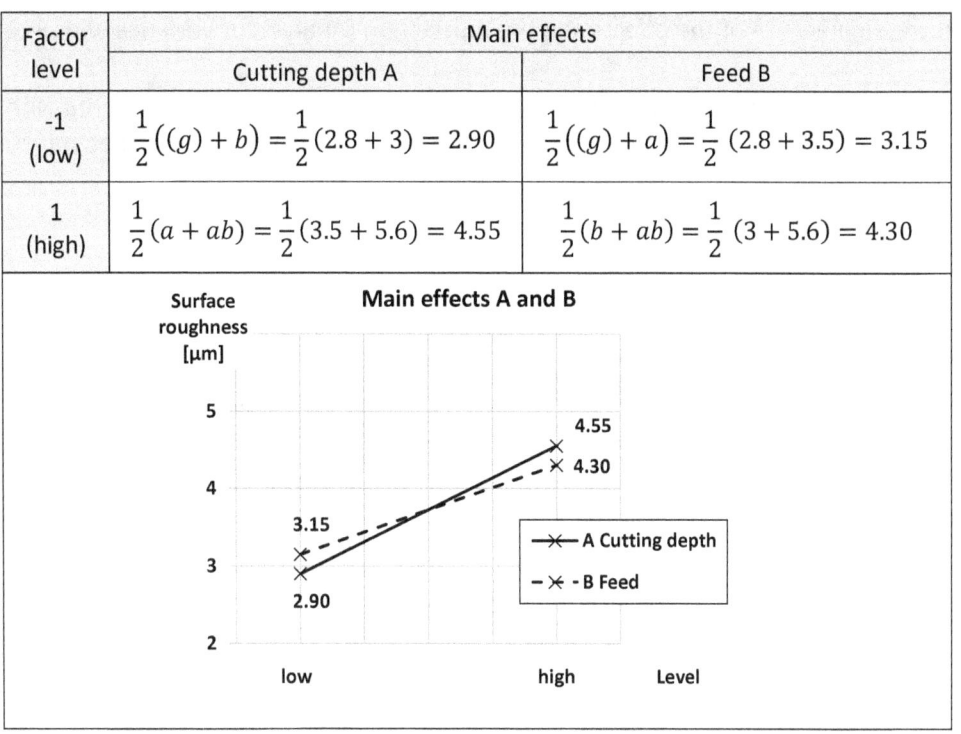

Table 25: Calculation of the point pairs and graphical representation of the main effects (main effects plot)

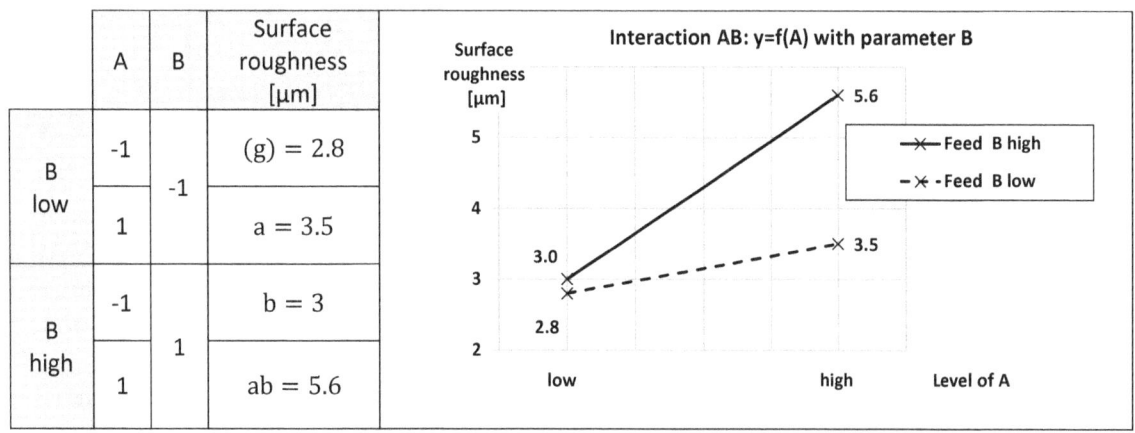

	A	B	Surface roughness [µm]
B low	-1	-1	$(g)=2.8$
	1		$a=3.5$
B high	-1	1	$b=3$
	1		$ab=5.6$

Table 26: Determination of the point pairs and graphical representation of the AB interaction (interaction plot); y=f(A) with parameter B

The strong interaction between feed and cutting depth can be seen with the aid of the diagram in Table 26: The cutting depth A has a greater effect on the surface roughness at higher feed rate than at a lower feed rate.

Column 4 of Table 27 contains the calculated F-values of the effects, which are compared with the critical value of the F-distribution. The table of the F-distribution, or a spreadsheet program or statistics software provides the following:

$$F_{1-\alpha}(1; f_R) = F_{0.95}(1; 4) \approx 7.7$$

Effects	Coefficients $b_0 = 3.725$	$MS_I = SS_I$	$F = \dfrac{MS_I}{MS_R}$	Significance
$A = 1.65$	$b_1 = 0.825$	$MS_A = A^2 = 2.7225$	$F_A = \dfrac{A^2}{MS_R} = \dfrac{2.7225}{0.105} \approx 25.9$	Significant, because $F_A > 7.7$
$B = 1.15$	$b_2 = 0.575$	$MS_B = B^2 = 1.3225$	$F_B = \dfrac{B^2}{MS_R} = \dfrac{1.3225}{0.105} \approx 12.6$	Significant, because $F_B > 7.7$
$AB = 0.95$	$b_3 = 0.475$	$MS_{AB} = (AB)^2 = 0.9025$	$F_{AB} = \dfrac{(AB)^2}{MS_R} = \dfrac{0.9025}{0.105} \approx 8.6$	Significant, because $F_{AB} > 7.7$

Alternative: Use the p-value to assess significance ($\alpha = 5\,\%$):

Effects	Coefficients $b_0 = 3.725$	$F = \dfrac{MS_I}{MS_R}$	p-value	Significance
$A = 1.65$	$b_1 = 0.825$	25.9	0.007	Significant, because $p < \alpha$
$B = 1.15$	$b_2 = 0.575$	12.6	0.024	Significant, because $p < \alpha$
$AB = 0.95$	$b_3 = 0.475$	8.6	0.043	Significant, because $p < \alpha$

Table 27: F-test to examine the significance of the effects and the interaction of the "Surface roughness of turned parts" 2^2 factorial design

The two main effects A and B, and the AB interaction are significant in this example because all their F-values are larger than the critical value $F_{0.95}(1; 4)$.

The predictive function is:

$$y = b_0 + b_1 x_A + b_2 x_B + b_3 x_A x_B$$

$$y = 3.725 \ \mu m + 0.825 x_A \ \mu m + 0.575 x_B \ \mu m + 0.475 x_A x_B \ \mu m$$

For the levels A^* and B^* demanded by the objective, the normalised quantities x_A and x_B have the following values:

$$x_A = \frac{A^* - \frac{1}{2}(A_2 + A_1)}{\frac{1}{2}(A_2 - A_1)} = \frac{0.8 - \frac{1}{2}(1 + 0.5)}{\frac{1}{2}(1 - 0.5)} = 0.20$$

$$x_B = \frac{B^* - \frac{1}{2}(B_2 + B_1)}{\frac{1}{2}(B_2 - B_1)} = \frac{0.04 - \frac{1}{2}(0.1 + 0.005)}{\frac{1}{2}(0.1 - 0.005)} \approx -0.263$$

The result for the predictive value is:

$$y = 3.725 \ \mu m + 0.825 \cdot 0.2 \ \mu m - 0.575 \cdot 0.263 \ \mu m - 0.475 \cdot 0.2 \cdot 0.263 \ \mu m \approx 3.7 \ \mu m$$

The result of the exemplary problem is: when the machine is set to a cutting depth of 0.8 mm and a feed rate of 0.04 mm per revolution, components with a surface roughness of 3.7 µm are to be expected.

2.5 Calculation model for the 2² factorial design

In the last chapter, a numerical example was used to show the step-by-step procedure for carrying out and evaluating a full factorial 2² factorial design with four runs. Below, a general calculation model is developed for the four steps shown. This can easily be done using a spreadsheet program. It is then a good idea to use this simple design to "play" in order to understand the relationships better[15]. The calculation model is then expanded further in the following chapters towards factorial designs with more than two factors. The latter govern the reality of DoE, of course. Their structure is principally based on the examples shown before[16].

1st step:	Plan and conduct experiments

The level combinations for the four experiments to be conducted are shown in Table 28. The last column is provided for entering the observations.

Run	Factor levels		Response y
	A	B	
(g)	-1	-1	$y_{(g)}$
a	1	-1	y_a
b	-1	1	y_b
ab	1	1	y_{ab}

Table 28: Schematic of the full factorial 2² experimental design

As has been said before, the setting of the factor levels is very important. Since runs require resources in terms of staff, time, equipment etc., the experimenter often faces a dilemma: on the one hand, they have to provide reliable results, and so there has to be sufficient data to carry out this task. On the other hand, they have to justify the experimental effort required for this. A general recommendation for setting the levels can therefore not be provided here.

---

[15] MS Excel®/OpenOffice Calc®-files for download, see Chapter 9.

[16] The considerations so far have discussed so-called full factorial designs, i.e. all factor combinations have been taken into account. Chapter 5 deals with the fact that, in reality, it is often necessary to switch to fractional factorial designs if there are more than two factors.

The effects and the interaction are calculated using the expressions in column 1 of Table 31. The *main effects plots* and the *interaction plots,* whose value pairs are determined in accordance with Table 29 or Table 30, serve to graphically illustrate the effects/interaction.

Factor level	Main effects	
	A	*B*
-1 (low)	$\frac{1}{2}((g) + b)$	$\frac{1}{2}((g) + a)$
1 (high)	$\frac{1}{2}(a + ab)$	$\frac{1}{2}(b + ab)$

Table 29: Calculation of the point pairs and graphical representation of the main effects

	Factor levels		y
	A	B	
B low	-1	-1	(g)
	1		a
B high	-1	1	b
	1		ab

Table 30: Determination of the point pairs for the graphical representation of the AB interaction (for the main effects plot y=f(A) with parameter B)

3rd step: Test the significance of the effects and the interaction (Analysis of Variance with F-test)

The sums of squares SS_R (as a measure for the experimental variance/error estimate) with degree of freedom f_R have to be known or be obtained from repeat runs and/or center point runs. For factorial designs with more factors (see next chapter), three-factor and multi-factor interactions exist which are rarely significant and can therefore be used to estimate the experimental variance. The mean squares are then calculated in accordance with:

$$MS_R = \frac{SS_R}{f_R}$$

The factorial design considered here with only four runs without repeats does not provide enough information to estimate the experimental variance, of course. This is assumed to be known. The F-test necessitates that the confidence level to be used for the test $1 - \alpha$ (usually: 95 %, rarely 90 %) be specified. Please note that how meaningful the F-test is strongly dependent on the number of values (= degrees of freedom f_R) . High critical values of the F-distribution are obtained particularly for factorial designs which are executed only once with two factors and therefore few values for the error estimate (few degrees of freedom). This makes it very difficult to find significant effects.

We would like to draw attention to the following chapters, where factorial designs (with repeat runs, too) with more than two factors are discussed.

Effects	Coefficients $b_0 = \bar{y}$	$MS_I = SS_I$	$F = \dfrac{MS_I}{MS_R}$	Effect significant if
$A = \dfrac{1}{2}(-(g) + a - b + ab)$	$b_1 = \dfrac{A}{2}$	A^2	$F_A = \dfrac{A^2}{MS_R}$	$F_A > F_{1-\alpha}(1; f_R)$
$B = \dfrac{1}{2}(-(g) - a + b + ab)$	$b_2 = \dfrac{B}{2}$	B^2	$F_B = \dfrac{B^2}{MS_R}$	$F_B > F_{1-\alpha}(1; f_R)$
$AB = \dfrac{1}{2}(+(g) - a - b + ab)$	$b_3 = \dfrac{AB}{2}$	$(AB)^2$	$F_{AB} = \dfrac{(AB)^2}{MS_R}$	$F_{AB} > F_{1-\alpha}(1; f_R)$

Table 31: Calculation model with Analysis of Variance for the 2^2 factorial design
(The significance of the effects can also be checked using the p-value:
They are significant when $p < \alpha$).

4th step: Draw up predictive function

$$y = \bar{y} + \frac{A}{2}x_A + \frac{B}{2}x_B + \frac{AB}{2}x_A x_B \quad \text{or}$$

$$y = b_0 + b_1 x_A + b_2 x_B + b_3 x_A x_B$$

For effects which are not significant, the corresponding coefficients b_1, b_2, b_3 are set to zero in these functional equations.

The following normalisation is used for the independent variables x_A and x_B of the predictive function:

$$x_A = \frac{A^* - \frac{1}{2}(A_2 + A_1)}{\frac{1}{2}(A_2 - A_1)} \qquad x_B = \frac{B^* - \frac{1}{2}(B_2 + B_1)}{\frac{1}{2}(B_2 - B_1)}$$

x_A and x_B are the minimum and maximum normalised values in the range of -1 to +1, respectively. A_1, B_1 and A_2, B_2 are the physical values of the low and the high levels of the factors.
A^* and B^* are the factor settings for which the response y to be expected is to be calculated.

3   The 2³ Factorial Design

The explanations and derivations which have been given so far have been explained with the aid of a 2^2 experimental design so they were easy to understand. With two factors, each set to two levels, there are *$2^2=4$* combinations - the minimum effort for this full factorial design thus consists of four runs.

In reality, it is very often the case that more than two factors exert an influence. This chapter now expands the methodology known from 2^2 factorial design to 2^3 factorial design: this means three factors which are each set to two levels. The number of runs - still without repetitions - is now *$2^3=8$.*

The following applies: 2^k ← Number of factors

↑
Number of levels per factor

You will see that the mathematics on which this is based is principally the same. The higher number of factor combinations makes it slightly more complex, however. The explanations which follow aim to develop 2^3 factorial design step by step from what is known about 2^2 factorial design. This will ultimately result in a methodology which enables designs, evaluations and result assessments of 2^3 factorial designs to be conducted according to a "recipe".

In the subsequent chapters, the concepts shown so far will then be developed further for more than three factors. You will recognise the familiar principles and thus be able to parameterise the DoE software as desired and use it with confidence. Using a DoE tool saves a lot of computational work in factorial design and in the numerical and graphical evaluation and the assessment of the observations.

3.1   Methodology and nomenclature

The designation of the runs and effects familiar from the previous chapters is now extended so that the mathematics for three factors can be derived. Figure 28 and Table 32 show the eight runs, which are represented by means of a cube. The rule for the designation of the runs applies here as well: factors at a high level are given by the corresponding small letters. Factors at low level are not given.

Example: For the run *ac*, the factors A and C are at a high level, while factor B is set at a low level.

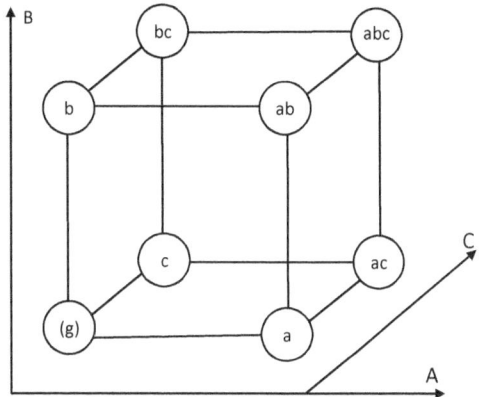

Figure 28: Designation of the runs of a 2³ factorial design.
(g) is the basic test, for which all factors are set at a low level.

Combinations of factor levels	$C_{low level}$		$C_{high level}$	
	$A_{low level}$	$A_{high level}$	$A_{low level}$	$A_{high level}$
$B_{low level}$	(g)	a	c	ac
$B_{high level}$	b	ab	bc	abc

Table 32: The eight combinations of the factor levels for 2³ factorial design

The designations of the runs in the standard order are (g), a, b, ab, c, ac, bc and abc.

The factors and their effects and interactions are designated by capital letters: A, B, AB, C, AC, BC and ABC. To distinguish between factors and effects in the text, the effects and interactions are written in *italics*.

Compared to the 2^2 factorial design, not only the main effect C but also the interactions AC and BC and the triple interaction ABC have been added.

In analogy with 2^2 factorial design, this factorial design is called a design or a model of first order: each factor here is set to two levels as well and the observations are used to determine the coefficients for a linear predictive function.

If one worked with three factor steps, the predictive function would describe a model of second order (quadratic model).

3.2 Effects and interactions

The definitions known from the 2^2 factorial design apply in general and thus here as well:

Main effect: Mean value of the differences in the observations when the factor was at a high or a low level.

Interaction: Mean of the differences in the effect of one factor at high level and the other factor at low level.

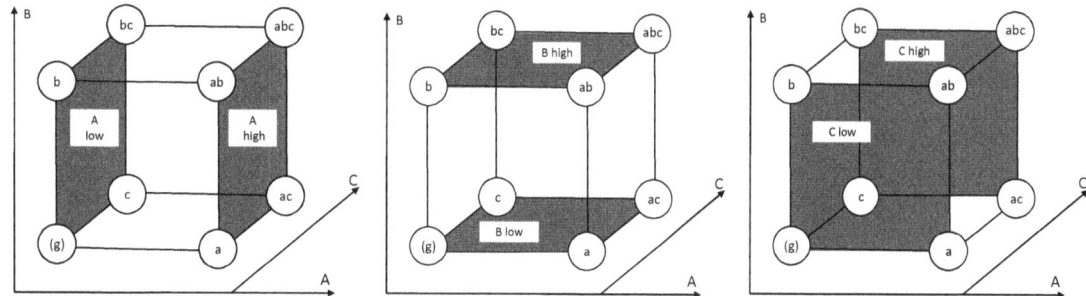

Figure 29: The graphical visualisation of the main effects A, B and C of a 2^3 factorial design in the cube model

With the aid of Figure 29, the calculation of the main effect *A* can be easily understood, for example:

Average of the measured values (responses) when A is at a <u>high</u> level:

$$A_{high} = \frac{1}{4}(a + ab + ac + abc)$$

Average of the measured values (responses) when A is at a <u>low</u> level:

$$A_{low} = \frac{1}{4}((g) + b + c + bc)$$

The effect of A as the difference in the averages of the measured values (responses):

$$A = A_{high} - A_{low} = \frac{1}{4}(a + ab + ac + abc) - \frac{1}{4}((g) + b + c + bc)$$

The methodology can also be recognised here: factor names which are stated (here a, ab, ac and abc) enter as positive values, all factor names which are not stated enter as negative values into the equation. The calculation rules for the other main effects can thus be written:

$$B = B_{high} - B_{low} = \frac{1}{4}(b + ab + bc + abc) - \frac{1}{4}((g) + a + c + ac)$$

$$C = C_{high} - C_{low} = \frac{1}{4}(c + ac + bc + abc) - \frac{1}{4}((g) + a + b + ab)$$

So how can the interactions be visualised for the 2^3 factorial design? This will be explained using the interaction between A and C by way of example. The task is to form the difference in the effect of A at high level of C and that at low level of C:

Effect of A at high level of C:

at high level of B:	$abc - bc$
at low level of B:	$ac - c$

Average of the sum: $A_{C\_high} = \frac{1}{4}(abc - bc) + \frac{1}{4}(ac - c)$

Effect of A at low level of C:

at high level of B:	$ab - b$
at low level of B:	$a - (g)$

Average of the sum: $A_{C\_low} = \frac{1}{4}(ab - b) + \frac{1}{4}(a - (g))$

The difference in these values is the *AC* interaction:

$$AC = A_{C\_high} - A_{C\_low} = \frac{1}{4}((g) + b + ac + abc) - \frac{1}{4}(a + ab + c + bc)$$

The interactions *AB* and *BC* can also be similarly calculated.

Sign schematic and multiplication rule

As already explained for the 2^2 factorial design, the sign schematic is also very helpful here for the run methodology and the calculation of the effects and interactions. Table 33 shows the methodology for the 2^3 factorial design. The runs are listed in the standard order *(g) ... abc* (1st column). The subsequent columns show the signs which are used to calculate the effects and interactions. The values in the columns with the interactions can be calculated by multiplying the values of the corresponding main effects (multiplication rule). The identity column I (all factors have a positive sign) is very useful for calculating the run average $\bar{y}$:

$$\bar{y} = \frac{1}{8}(+(g) + a + b + ab + c + ac + bc + abc)$$

Run	Signs of effects							
	I	A	B	AB	C	AC	BC	ABC
(g)	+	-	-	+	-	+	+	-
a	+	+	-	-	-	-	+	+
b	+	-	+	-	-	+	-	+
ab	+	+	+	+	-	-	-	-
c	+	-	-	+	+	-	-	+
ac	+	+	-	-	+	+	-	-
bc	+	-	+	-	+	-	+	-
abc	+	+	+	+	+	+	+	+

Table 33: The sign schematic for the effects and interactions for the 2^3 factorial design. For clarity, the levels have been denoted by +/ - instead of 1/-1.

With Table 33, the double interactions are now as follows:

$$AB = \frac{1}{4}(+(g) - a - b + ab + c - ac - bc + abc) = \frac{1}{4}((g) + ab + c + abc) - \frac{1}{4}(a + b + ac + bc)$$

$$AC = \frac{1}{4}(+(g) - a + b - ab - c + ac - bc + abc) = \frac{1}{4}((g) + b + ac + abc) - \frac{1}{4}(a + ab + c + bc)$$

$$BC = \frac{1}{4}(+(g) + a - b - ab - c - ac + bc + abc) = \frac{1}{4}((g) + a + bc + abc) - \frac{1}{4}(b + ab + c + ac)$$

The calculation rule for the triple interaction *ABC* can similarly be determined using the schematic by multiplying the signs of the columns *A*, *B* and *C*: graphically, the *ABC* interaction can be found as the difference between the *AB* interaction at high level of *C* and that at low level of *C*.

$$ABC = AB_{C\_high} - AB_{C\_low}$$

Interaction *AB* at high level of C:

at high level of B:	$abc - bc$
at low level of B:	$ac - c$
Average of the difference:	$AB_{C\_high} = \frac{1}{4}(abc - bc) - \frac{1}{4}(ac - c)$

Interaction AB at low level of C:

at high level of B:	$ab - b$
at low level of B:	$a - (g)$

Average of the difference: $\quad ABC_{\_low} = \frac{1}{4}(ab - b) - \frac{1}{4}(a - (g))$

The difference in these values gives the ABC interaction:

$$ABC = ABC_{\_high} - ABC_{\_low} = \frac{1}{4}(a + b + c + abc) - \frac{1}{4}((g) + ab + ac + bc)$$

The calculation of the effects and interactions will now be shown using an example.

Example: Product yield of a chemical reactor (3 factors, 2 levels)

The effect of three factors on the amount of a final product which the reaction in a chemical reactor produces per unit of time is investigated. The three factors which (presumably) affect this yield are the reactor temperature A, the concentration B of a particular component of the formulation, and the design C of the catalytic system used.

The runs were conducted at the levels of the factors stated in Table 34.

Response: Product yield [kg]				Factor levels	
	Factor	Measurement unit	Factor type	Low	High
A	Temperature	°C	Quantitative	160	180
B	Concentration	%	Quantitative	20	40
C	Catalytic system	Dimensionless	Qualitative	Cat X	Cat Y

Table 34: The levels of the factors for the 2^3 factorial design
(Observations see Table 35)

The catalytic system (factor C) is a qualitative (dimensionless) factor. Cat X and Cat Y are specific types of catalytic system, for example similar designs from different manufacturers. There are no values between the two levels for qualitative factors.

The observations of the eight runs are shown in Table 35. The response is the quantity of end product in kg which is produced in a specified period of time.

Concentration B		Catalytic system C							
		Cat X	-			Cat Y	+		
		Temperature A							
		160	-	180	+	160	-	180	+
20	-	(g)	68	a	82	c	59	ac	94
40	+	b	61	ab	77	bc	51	abc	91

Table 35: The observations of the "Product yield of a chemical reactor" 2^3 factorial design

The effects are now calculated using the sign schematic (Table 33). This will be shown again using the *AB* interaction by way of example:

$$AB = \frac{1}{4}(+(g) - a - b + ab + c - ac - bc + abc) = \frac{1}{4}(68 - 82 - 61 + 77 + 59 - 94 - 51 + 91) = 1.75$$

The interaction, which is positive here, thus has the value 1.75 kg.

The further effects and interactions, whose results are summarised in Table 36, are calculated in a similar way.

Effect	
Name	Value [kg]
A	26.25
B	-5.75
AB	1.75
C	1.75
AC	11.25
BC	0.25
ABC	0.75

Table 36: The calculated effects and interactions of the "Product yield of a chemical reactor" 2^3 factorial design

3.2.1 Graphical representation of the effects and interactions

It is possible to draw up an illustrative representation of the effects calculated in the previous chapter. As shown for the 2^2 factorial design, two averages, which are formed from the observations at low and high level of the corresponding factor, are connected in the diagram for each of the three main effects. The calculation of the averages is shown in Table 37.

Factor level	Factor A	Factor B	Factor C
	Temperature	Concentration	Catalytic system
-	$\frac{1}{4}((g)+b+c+bc)=$ $\frac{1}{4}(68+61+59+51)=$ 59.75	$\frac{1}{4}((g)+a+c+ac)=$ $\frac{1}{4}(68+82+59+94)=$ 75.75	$\frac{1}{4}((g)+a+b+ab)=$ $\frac{1}{4}(68+82+61+77)=$ 72.00
+	$\frac{1}{4}(a+ab+ac+abc)=$ $\frac{1}{4}(82+77+94+91)=$ 86.00	$\frac{1}{4}(b+ab+bc+abc)=$ $\frac{1}{4}(61+77+51+91)=$ 70.00	$\frac{1}{4}(c+ac+bc+abc)=$ $\frac{1}{4}(59+94+51+91)=$ 73.75

Table 37: The averages of the observations (each at low and high level of the factors) form the starting and end points of the three straight lines for the main effects.

Figure 30 shows the graphical representation of the main effects.

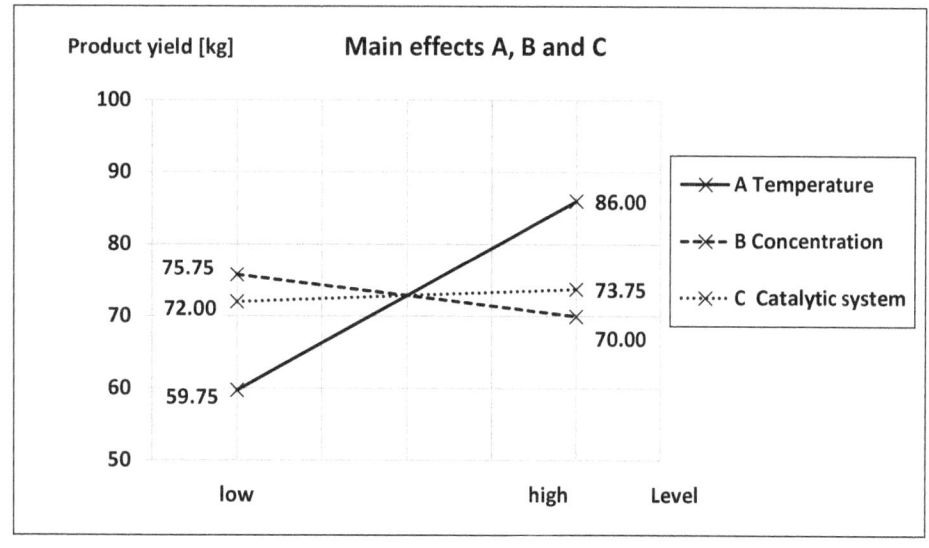

Figure 30: The graphical representation of the main effects (Main effects plot)

The procedure for the three interactions *AB*, *AC* and *BC* of the 2^3 factorial design is as follows: the two averages of the responses of all minus and plus settings of the one factor are calculated as a function of the settings of the other factor. This results in two pairs of points per interaction, which are shown in the interaction plot (Figure 31 to Figure 33).

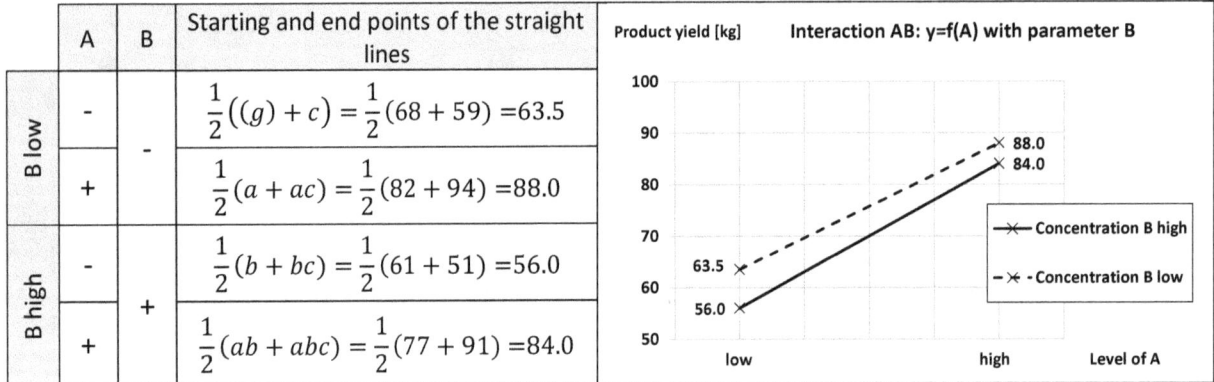

	A	B	Starting and end points of the straight lines	
B low	-	-	$\frac{1}{2}((g)+c) = \frac{1}{2}(68+59) = 63.5$	
B low	+	-	$\frac{1}{2}(a+ac) = \frac{1}{2}(82+94) = 88.0$	
B high	-	+	$\frac{1}{2}(b+bc) = \frac{1}{2}(61+51) = 56.0$	
B high	+	+	$\frac{1}{2}(ab+abc) = \frac{1}{2}(77+91) = 84.0$	

Figure 31: AB interaction, y=f(A) with parameter B

	A	C	Starting and end points of the straight lines	
C low	-	-	$\frac{1}{2}((g)+b) = \frac{1}{2}(68+61) = 64.5$	
C low	+	-	$\frac{1}{2}(a+ab) = \frac{1}{2}(82+77) = 79.5$	
C high	-	+	$\frac{1}{2}(c+bc) = \frac{1}{2}(59+51) = 55.0$	
C high	+	+	$\frac{1}{2}(ac+abc) = \frac{1}{2}(94+91) = 92.5$	

Figure 32: AC interaction, y=f(A) with parameter C

	B	C	Starting and end points of the straight lines	
C low	-	-	$\dfrac{1}{2}((g)+a)=\dfrac{1}{2}(68+82)=75.0$	
	+		$\dfrac{1}{2}(b+ab)=\dfrac{1}{2}(61+77)=69.0$	
C high	-	+	$\dfrac{1}{2}(c+ac)=\dfrac{1}{2}(59+94)=76.5$	
	+		$\dfrac{1}{2}(bc+abc)=\dfrac{1}{2}(51+91)=71.0$	

Figure 33: BC interaction, y=f(B) with parameter C

As has already been explained for the 2^2 design, the degree of "non-parallelism" is a measure of the "strength" of the interaction. If the straight lines run parallel, there is no interaction.

Note that multiple interactions are usually not significant in practice, however. The consideration of the triple interaction *ABC* has therefore been omitted here. For the subsequent F-test to determine the significance of the other effects, the result of the run abc is used to estimate the experimental variance.

3.3 Significance of effects: Analysis of Variance (F-test)

As has already been done for the 2^2 factorial design, the significance of effects and interactions is to be assessed here as well by means of an Analysis of Variance with F-test.

The objective of the Analysis of Variance is to decide whether each effect and each interaction is significant. To this end, each test value F is formed as shown below in order to subsequently be compared with the critical value of the F-distribution:

$$F = \frac{\frac{SS_I}{f_I}}{\frac{SS_R}{f_R}} = \frac{MS_I}{MS_R}$$

The observations are again divided into two groups. Group 1 contains the values where the factor investigated was at a low level. Group 2 contains the values at the high factor levels. Table 38 shows the groups for the Analysis of Variance using the example of factor A.

$\dfrac{j}{i}$	1	2	3	4	Group sum	Group mean $\bar{y}_i$
Group 1: A_{low}	(g)	b	c	bc	$(g) + b + c + bc$	$\dfrac{(g) + b + c + bc}{4}$
Group 2: A_{high}	a	ab	ac	abc	$a + ab + ac + abc$	$\dfrac{a + ab + ac + abc}{4}$

Table 38: The two groups of the Analysis of Variance for factor A of a 2^3 factorial design

The following applies for the Analysis of Variance:

Group index:	i
Value index in group:	j
Number of groups:	$I = 2$
Degrees of freedom:	$f_I = I - 1 = 1$
Number of values per group:	$J = 4$
Group mean value:	$\bar{y}_i$
Total mean value:	$\bar{y}$

The sum of squares (between the groups) for the factor A is calculated as follows:

$$SS_I = SS_A = J \cdot \sum_{i=1}^{I} (\bar{y}_i - \bar{y})^2 =$$

$$4\left(\frac{(g)+b+c+bc}{4} - \frac{(g)+b+c+bc+a+ab+ac+abc}{8}\right)^2 +$$

$$4\left(\frac{a+ab+ac+abc}{4} - \frac{(g)+b+c+bc+a+ab+ac+abc}{8}\right)^2 =$$

$$4\left(\frac{(g)+b+c+bc-(a+ab+ac+abc)}{8}\right)^2 + 4\left(\frac{a+ab+ac+abc-((g)+b+c+bc)}{8}\right)^2 =$$

$$4\left(\frac{-4A}{8}\right)^2 + 4\left(\frac{+4A}{8}\right)^2 = 2A^2$$

This relationship between the effects and the sums of squares applies for all effects and interactions:

$$SS_A = 2A^2$$
$$SS_B = 2B^2$$
$$SS_{AB} = 2(AB)^2$$
$$SS_C = 2C^2$$
$$SS_{AC} = 2(AC)^2$$
$$SS_{BC} = 2(BC)^2$$
$$SS_{ABC} = 2(ABC)^2$$

The methodology for calculating the sums of squares from the effects is:

$$SS_{Effect} = \frac{N}{4} \cdot (Effect)^2 \qquad \text{(N = number of runs).}$$

The mean squares are calculated according to $MS_I = \frac{SS_I}{f_I}$.

Inserting $f_I = I - 1 = 1$ gives:

$$MS_A = \frac{SS_A}{1} = SS_A$$
$$MS_B = SS_B$$
$$\dots$$
$$MS_{ABC} = SS_{ABC}$$

Inserting the numerical values of the example now gives the mean squares:

$$MS_A = 2A^2 = 2 \cdot 26.25^2 = 1\,378.125$$
$$MS_B = 2B^2 = 2 \cdot (-5.75)^2 = 66.125$$

...

$$MS_{ABC} = 2(ABC)^2 = 2 \cdot 0.75^2 = 1.125$$

The last column of Table 39 shows the mean squares for all effects:

Effect		Mean squares
Name	Value [kg]	$SS_I = MS_I$
A	26.25	1 378.125
B	-5.75	66.125
AB	1.75	6.125
C	1.75	6.125
AC	11.25	253.125
BC	0.25	0.125
ABC	0.75	1.125

Table 39: Average sums of the mean squares calculated from the effects

MS_R is now required as a measure of the experimental error. In practice, this value is often known from previous experiments. In the present example, it can be seen that the sums of squares SS_I of the interactions BC and ABC have very low values compared to the main effects and the AC interaction. This results in very low F values for these effects, which then leads to the test result "effect is not significant". The low values of the mean squares obtained originate in all probability from the scatter of the observations. They are therefore frequently used to estimate the experimental variance:

$$SS_R = SS_{BC} + SS_{ABC}$$

The following applies to the degree of freedom: $f_R = f_{BC} + f_{ABC} = 1 + 1 = 2$

The mean square sought is thus:

$$MS_R = \frac{SS_{BC} + SS_{ABC}}{f_R} = \frac{0.125 + 1.125}{2} = 0.625$$

The F-test states that an effect under investigation is significant when the following applies:

$$F = \frac{MS_I}{MS_R} > F_{1-\alpha}(f_I; f_R)$$

For the present example, the following applies for effect A:

$$F_A = \frac{MS_A}{MS_R} = \frac{1\,378.125}{0.625} = 2\,205.0$$

If the F-test is conducted at a confidence level of 95 %, the critical value of the F-distribution is:

$$F_{1-\alpha}(f_I; f_R) = F_{0.95}(1; 2) \approx 18.51$$

The result of the F-test for effect A is: Since the F value calculated (2 205.0) is greater than the critical value of the F-distribution (18.51), it is assumed with a significance level of 5 % that effect A is significant.

Note that in the present case, the experimental variance MS_R was estimated with the aid of only two values. The data basis for the decision as to the significance is therefore rather "thin". Here as well, greater certainty is obtained by having a larger number of runs, for example by repeating runs several times.

The F-tests for the further effects and interactions are now conducted in this way. The results are shown in Table 40.

Effect					
Name	Value [kg]	$SS_I = MS_I$	$F = MS_I/MS_R$	p-value	F-test result
A	26.25	1 378.125	2 205.00	0.0005	Significant
B	-5.75	66.125	105.80	0.0093	Significant
AB	1.75	6.125	9.80	0.0887	
C	1.75	6.125	9.80	0.0887	
AC	11.25	253.125	405.00	0.0025	Significant
BC	0.25	0.125	0.20	0.6985	
ABC	0.75	1.125	1.80	0.3118	

Table 40: The last column contains the results of the F-tests: The effects A, B and the AC interaction are significant

As a result of this Analysis of Variance, we note that the main effects A and B, and the AC interaction are identified as being significant. One more consideration is worth mentioning here, however: how can the AC interaction be significant, when the main affect C was assessed as being not significant? The suspicion could be that the main effect C is hidden, i. e. that the assumed linear influence of factor C on the response does not exist. As shown in Chapter 2.3.2, this assumption can be confirmed or contradicted by center point runs. A quadratic model must then be used for the predictive function, where necessary.

A further comment on the estimation of the variance (experimental error): the method used in the explanations above, which summarises (possibly) not significant interactions to calculate the experimental error, is also called "pooling". Which values are used for the pooling is at the discretion of the experimenter and is therefore somewhat subjective. The larger the number of values used for the pooling, the less critical this subjectivity becomes, however. Here as well, center point runs and/or repeat runs can improve the degree of certainty regarding the significance.

3.4 Predictive function

The general predictive function for the 2^3 factorial design (model of first order) is:

$$y = \bar{y} + \frac{A}{2}x_A + \frac{B}{2}x_B + \frac{AB}{2}x_A x_B + \frac{C}{2}x_C + \frac{AC}{2}x_A x_C + \frac{BC}{2}x_B x_C + \frac{ABC}{2}x_A x_B x_C$$

Or written using the coefficients b_0 to b_7:

$$y = b_0 + b_1 x_A + b_2 x_B + b_3 x_A x_B + b_4 x_C + b_5 x_A x_C + b_6 x_B x_C + b_7 x_A x_B x_C$$

Table 41 lists the calculated coefficients of the numerical example.

Effect		Coefficients b_0 to b_7	$SS_I=MS_I$	$F=MS_I/MS_R$	p-value	F-test result
Name	Value [kg]	72.875				
A	26.25	13.125	1 378.125	2 205.00	0.0005	Significant
B	-5.75	-2.875	66.125	105.80	0.0093	Significant
AB	1.75	0.875	6.125	9.80	0.0887	
C	1.75	0.875	6.125	9.80	0.0887	
AC	11.25	5.625	253.125	405.00	0.0025	Significant
BC	0.25	0.125	0.125	0.20	0.6985	
ABC	0.75	0.375	1.125	1.80	0.3118	

Table 41: The calculated effects, coefficients of the predictive function, and the results of the F-tests for the 2^3 factorial design "Product yield of a chemical reactor"

The mathematical model is intended to represent a linear relationship between the factors and the response y with the values found through the experiments, after all. For a rough check of the plausibility of the predictive function, the standard values $x_A = x_B = x_C = 1$ can be set, for example. One then obtains the calculated observation y for the case that all three factors are at a high level:

$$y = b_0 + b_1 + b_2 + b_3 + b_4 + b_5 + b_6 + b_7$$

With the numerical values of the last chapter, one obtains:

$$y = (\,72.875 + 13.125 - 2.875 + 0.875 + 0.875 + 5.625 + 0.125 + 0.375)\ \text{kg} = 91\ \text{kg}$$

The value of 91 kg obtained corresponds to the result of the run abc.

Since the Analysis of Variance in the last chapter showed that not all effects are significant, the equation has to be adapted accordingly. This is done by setting the effect values of the effects which are not significant to 0.

This means: $AB = C = BC = ABC = 0$

The following predictive function is thus obtained:

$$y = b_0 + b_1 x_A + b_2 x_B + b_5 x_A x_C$$

$$y = (72.875 + 13.125 \cdot x_A - 2.875 \cdot x_B + 5.625 \cdot x_A \cdot x_C) \text{ kg}$$

By way of example, it will now be used to calculate the response which would result from the following factor settings:

Temperature A^*: 177°C
Concentration B^*: 35 %
Catalytic system C^*: CAT Y

First of all, those transformations are repeated which set the experimental space for each factor from -1 to +1. The following dimensionless quantities are obtained:

$$x_A = \frac{A^* - \frac{1}{2}(A_2 + A_1)}{\frac{1}{2}(A_2 - A_1)} = \frac{177 - \frac{1}{2}(180 + 160)}{\frac{1}{2}(180 - 160)} = 0.70$$

$$x_B = \frac{B^* - \frac{1}{2}(B_2 + B_1)}{\frac{1}{2}(B_2 - B_1)} = \frac{35 - \frac{1}{2}(40 + 20)}{\frac{1}{2}(40 - 20)} = 0.50$$

In the case of factor C, it is a qualitative factor where no values between the levels can be set. In the present task, the catalytic system CAT Y is to be used. This corresponds to factor C at high level. We therefore have to write: $x_C = 1$.
When inserted into the predictive function, this results in the following predictive value for the levels assumed:

$$y = (72.875 + 13.125 \cdot 0.7 - 2.875 \cdot 0.5 + 5.625 \cdot 0.7 \cdot 1) \text{ kg} \approx 84.6 \text{ kg}$$

With the setting stated, this would result in a product yield of 84.6 kg.

The considerations from this and the previous chapters are repeated below with the aid of examples.

3.5 2^3 example "Adhesive strength of a bond"

The 2^3 examples calculated in this and the subsequent chapters show the practical procedure step by step. This is then used to develop the calculation methodology for full factorial 2^3 factorial designs. The very simple arithmetic can easily be implemented using a spreadsheet program (download of the examples in the book see Appendix 9).

Task: Adhesive strength of a bond

A factorial design was drawn up to investigate the adhesive strength of an industrial adhesive. The factors are: how thickly the adhesive has been applied (coating thickness), the contact pressure and the length of time pressure is applied to the adhesive on two test surfaces to be bonded (contact duration). The F-test as part of the Analysis of Variance to assess the significance of the effects shall be based on a significance level of 5 %. The task is to use the predictive function to calculate which adhesive strength is to be expected when the work is conducted with a coating thickness of 38 g/m², a contact pressure of 12 N/cm² and a contact duration of 20 hours.[17]

1st step: Plan and conduct experiments

After the levels of the three factors have been fixed or specified, the eight runs are carried out with the corresponding level combinations (Table 42 and Table 43).

	Response: Adhesive strength [kN]			Factor levels	
	Factor	Measurement unit	Factor type	Low	High
A	Coating thickness	g/m²	Quantitative	30	40
B	Contact pressure	N/cm²	Quantitative	10	20
C	Contact duration	h	Quantitative	1	24

Table 42: Levels of the 2^3 factorial design "Adhesive strength of a bond"

Contact pressure B		Contact duration C							
		1	-			24	+		
		Coating thickness A							
		30	-	40	+	30	-	40	+
10	-	(g)	30.4	a	34.1	c	33.9	ac	38.4
20	+	b	32.1	ab	32.9	bc	37.3	abc	38.4

Table 43: Observations of the 2^3 factorial design "Adhesive strength of a bond"

---

[17] Following Adam, Mario: Statistische Versuchsplanung und Auswertung

The effects and the interactions are calculated in accordance with Table 37. The coefficients of the predictive function are calculated from the effects by dividing them by two (Table 44).

Effect		Coefficients b_0 to b_7
Name	Value [kN]	
		34.688
A	2.525	1.263
B	0.975	0.488
AB	-1.575	-0.788
C	4.625	2.313
AC	0.275	0.138
BC	0.725	0.363
ABC	-0.125	-0.062

Table 44: Effects, interactions and coefficients of the predictive function of the 2³ factorial design "Adhesive strength of a bond"

The start and end points for the main effects plot can be derived with the aid of the equations of Figure 31 to Figure 33. The corresponding diagrams are summarised in Figure 34.

The diagram illustrates the *AB* interaction between coating thickness and contact pressure very clearly: the increase in coating thickness A has a stronger effect at low contact pressure B than at high contact pressure.

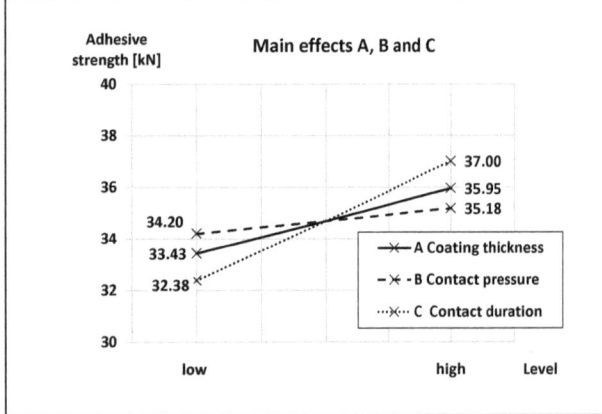
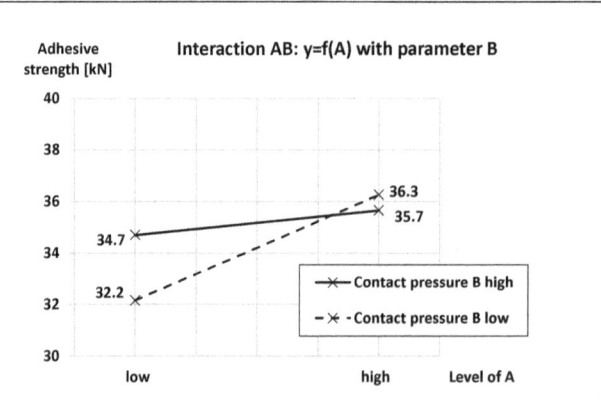

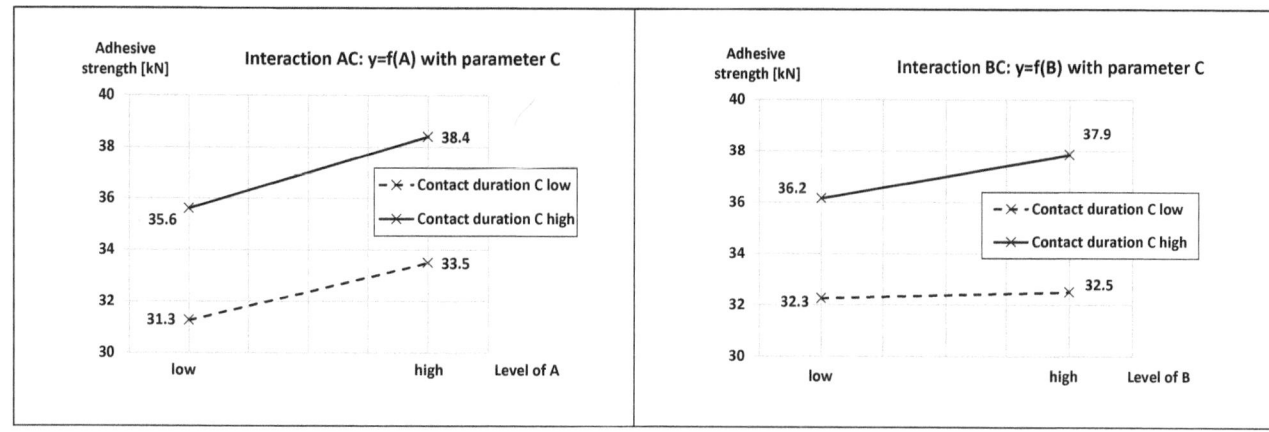

Figure 34: Graphical representation of the effects and the two-fold interactions of the 2^3 factorial design "Adhesive strength of a bond"

3rd step: Test the significance of the effects and interactions (Analysis of Variance with F-test)

Since the sums of squares SS_I of the interactions AC, BC and ABC have very low values (Table 45), they are used to estimate the variance.

$$SS_R = SS_{AC} + SS_{BC} + SS_{ABC}$$

The degree of freedom is $f_R = f_{AC} + f_{BC} + f_{ABC} = 1 + 1 + 1 = 3$

The mean square is thus:

$$MS_R = \frac{SS_{AC} + SS_{BC} + SS_{ABC}}{f_R} \approx \frac{0.151 + 1.051 + 0.031}{3} \approx 0.411$$

Column 5 of Table 45 contains the F-values of the effects calculated with this value. From the table of the F-distribution, a spreadsheet program or statistics software, the following critical value of the F-distribution at the specified confidence level of 95 % is obtained:

$$F_{1-\alpha}(1; f_R) = F_{0.95}(1; 3) \approx 10.13$$

The result of the test is: the two main effects A and C, and the AB interaction are significant because all their F-values are larger than the critical value $F_{0.95}(1; 3)$. Since the AB interaction was identified as being significant and the associated main effect B was identified as being not significant, further runs are necessary to assess whether B is a hidden effect (see 2.3.2).

Cases in which higher order interactions (for example ABC) are identified as being significant are similar. Associated lower order interactions (for example AB) and also main effects (for example C) must then not be used for the estimation of the variance.

Effect		Coeff. b_0 to b_7	$SS_i=MS_i$	$F=MS_i/MS_R$	p-value	F-test result
Name	Value [kN]					
		34.688				
A	2.525	1.263	12.751	31.01	0.011	Significant
B	0.975	0.488	1.901	4.62	0.121	
AB	-1.575	-0.788	4.961	12.06	0.040	Significant
C	4.625	2.313	42.781	104.03	0.002	Significant
AC	0.275	0.138	0.151	0.37	0.587	
BC	0.725	0.363	1.051	2.56	0.208	
ABC	-0.125	-0.062	0.031	0.08	0.801	

Table 45: F-test to examine the significance of the effects and interactions of the 2^3 factorial design "Adhesive strength of a bond"

4th step: Draw up predictive function

In the predictive function

$$y = b_0 + b_1 x_A + b_2 x_B + b_3 x_A x_B + b_4 x_C + b_5 x_A x_C + b_6 x_B x_C + b_7 x_A x_B x_C$$

the coefficients of the effects which are not significant are set to zero. One thus obtains:

$$y = (34.688 + 1.263 \cdot x_A - 0.788 \cdot x_A \cdot x_B + 2.313 \cdot x_C) \, kN$$

For the calculation of the normalised quantities x_A, x_B and x_C for the levels A^*, B^* and C^* specified in the task, the following are obtained:

$$x_A = \frac{A^* - \frac{1}{2}(A_2 + A_1)}{\frac{1}{2}(A_2 - A_1)} = \frac{38 - \frac{1}{2}(40 + 30)}{\frac{1}{2}(40 - 30)} = 0.6$$

$$x_B = \frac{B^* - \frac{1}{2}(B_2 + B_1)}{\frac{1}{2}(B_2 - B_1)} = \frac{12 - \frac{1}{2}(20 + 10)}{\frac{1}{2}(20 - 10)} = -0.6$$

$$x_C = \frac{C^* - \frac{1}{2}(C_2 + C_1)}{\frac{1}{2}(C_2 - C_1)} = \frac{20 - \frac{1}{2}(24 + 1)}{\frac{1}{2}(24 - 1)} \approx 0.652$$

The result for the predictive value is:

$$y = (34.688 + 1.263 \cdot 0.6 - 0.788 \cdot 0.6 \cdot (-0.6) + 2.313 \cdot 0.652) \, kN \approx 37.2 \, kN.$$

The result is: for 38 g/m² of adhesive, a contact pressure of 12 N/cm² and a contact duration of 20 hours, an adhesive strength of 37.2 kN between the test surfaces is to be expected.

3.6 2^3 example "Processing time of an offer"

This example shows how a commercial process can be investigated and optimised with the aid of a factorial design with three qualitative factors.

Task: Processing time of an offer

A company has identified that the time needed to process offers is too long and that orders are lost because of this. The suspected causes of the problem are:

- The number of organisational interfaces during offer processing is too high.
- The work of the planning and purchasing organisational units is not simultaneous enough.
- The selection of suitable subsuppliers takes too long.

The following suggestions for a solution are worked out for the three suspected causes:

- Reduce the division of work as the offer is being drawn up. The offer then goes through fewer "pairs of hands".
- Make work stages as parallel as possible.
- Reduce the number of potential suppliers in order to reduce the effort involved in selecting suppliers.

The procedure has four steps as usual:

1st step: Plan and conduct experiments

Table 46 shows the factors and the levels to be set for them.

	Response: Processing time [days]			Factor levels	
	Factor	Measurement unit	Factor type	Low	High
A	Interfaces	-	Qualitative	Reduced	As before
B	Simultaneous operation	-	Qualitative	Increased	As before
C	Supplier selection	-	Qualitative	Faster	As before

Table 46: Levels of the 2^3 factorial design "Processing time of an offer"

All three factors in this example are qualitative in nature. The evaluation of the observations aims primarily to determine the significance of the effects in order to initiate the correct measures to improve the process, where necessary. A predictive function can be helpful in this case when the factors can also have intermediate values.

The observations can be seen in Table 47.

Run	Processing time [days]
(g)	4.6
a	6.2
b	7.8
ab	9.4
c	5.1
ac	6.4
bc	8.4
abc	10.2

Table 47: Observations of the 2^3 factorial design "Processing time of an offer"

2nd step: Calculation and graphical representation of the effects and interactions

The effects, the interactions and the coefficients of the predictive function are calculated using the equations known from the previous chapters (see also Table 53). The results are shown in Table 48.

Effect		Coeff. b_0 to b_7
Name	Value [days]	
		7.263
A	1.575	0.788
B	3.375	1.688
AB	0.125	0.062
C	0.525	0.263
AC	-0.025	-0.013
BC	0.175	0.088
ABC	0.125	0.062

Table 48: Calculation of the effects, interactions and coefficients of the predictive function of the 2^3 factorial design "Processing time of an offer"

The pairs of points for the graphical representations of the effects and the two-fold interactions are determined according to the known methodology (see alsoTable 52). The diagrams are summarised in Figure 35. Since the straight lines are largely parallel, the three diagrams for the interactions suggest that *AB*, *AC* and *BC* are not significant. This is to be statistically proven in step 3 by means of the Analysis of Variance with F-test.

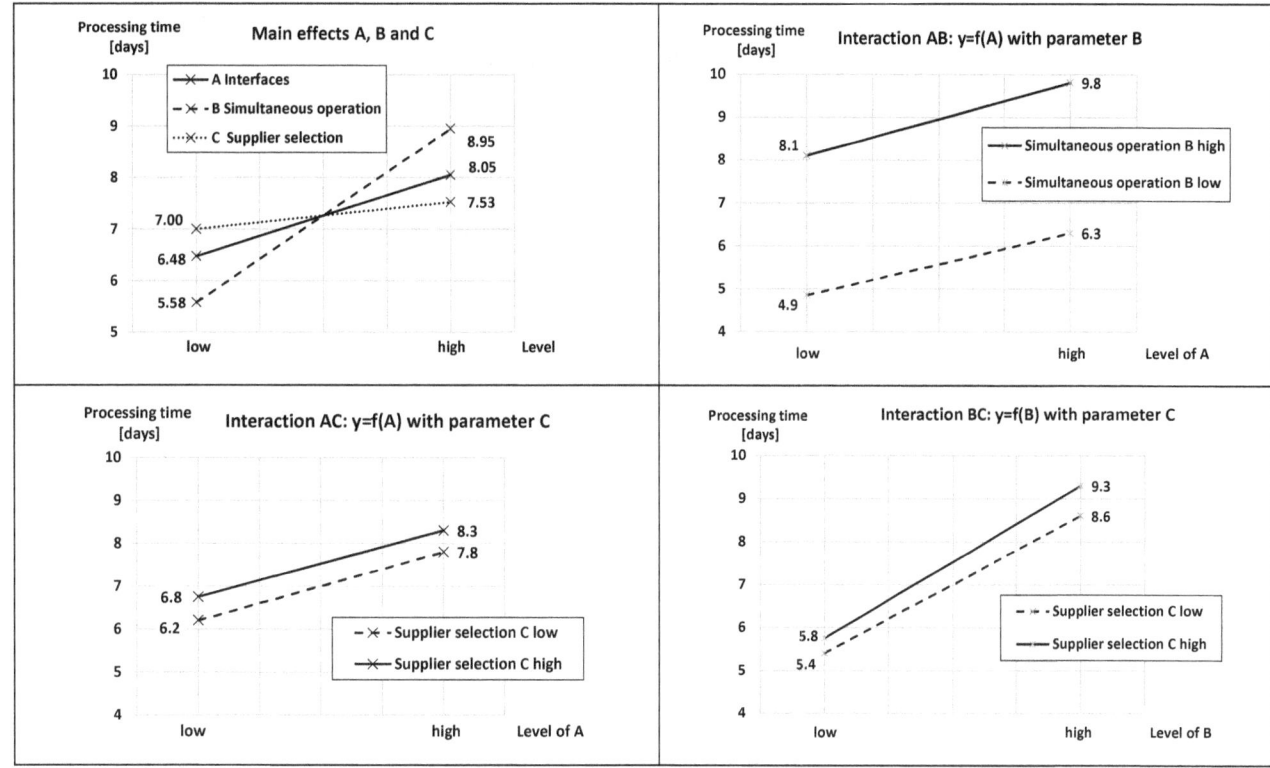

Figure 35: Graphical representation of the effects and the two-fold interactions of the
2^3 factorial design "Processing time of an offer"

3rd step: Test the significance of the effects and interactions (Analysis of Variance with F-test)

Since the sums of squares SS_I of all interactions have very low values compared to the main effects (Table 49), they are used for the estimation of the variance:

$$SS_R = SS_{AB} + SS_{AC} + SS_{BC} + SS_{ABC}$$

The degree of freedom is $f_R = f_{AB} + f_{AC} + f_{BC} + f_{ABC} = 1 + 1 + 1 + 1 = 4$

The mean square is thus:

$$MS_R = \frac{SS_{AB} + SS_{AC} + SS_{BC} + SS_{ABC}}{f_R} \approx \frac{0.031 + 0.001 + 0.061 + 0.031}{4} \approx 0.0313$$

Table 49 contains the F-values of the effects calculated with this value. From the table of the F-distribution, a spreadsheet program or statistics software, the following critical value of the F-distribution at a confidence level of 95 % is obtained:

$$F_{1-\alpha}(1; f_R) = F_{0.95}(1; 4) \approx 7.71$$

The result of the test is: all three main effects are significant. The effect B (planning and purchasing work in parallel) is most important here. The simultaneous work in these two organisational units has the greatest effect on reducing the processing times of offers.

Effect		Coeff. b_0 to b_7	$SS_i = MS_i$	$F = MS_i / MS_R$	p-value	F-test result
Name	Value [days]	7.263				
A	1.575	0.788	4.961	158.76	0.00023	Significant
B	3.375	1.688	22.781	729.00	0.00001	Significant
AB	0.125	0.062	0.031	1.00	0.37390	
C	0.525	0.263	0.551	17.64	0.01370	Significant
AC	-0.025	-0.013	0.001	0.04	0.85124	
BC	0.175	0.088	0.061	1.96	0.23410	
ABC	0.125	0.062	0.031	1.00	0.37390	

Table 49: Significance of the effects and interactions of the 2^3 factorial design "Processing time of an offer"

4th step: Draw up predictive function

In the predictive function

$$y = b_0 + b_1 x_A + b_2 x_B + b_3 x_A x_B + b_4 x_C + b_5 x_A x_C + b_6 x_B x_C + b_7 x_A x_B x_C$$

the coefficients of the effects which are not significant are set to zero. One thus obtains:

$$y = (7.263 + 0.788 \cdot x_A + 1.688 \cdot x_B + 0.263 \cdot x_C)\ days$$

According to the task set, the processing time is to be calculated for the case that all three factors are at a low level. This corresponds to the implementation of the three proposed improvement measures. With the normalised quantities $x_A = x_B = x_C = -1$, the predictive function results in:

$$y = (7.263 - 0.788 - 1.688 - 0.263)\ days \approx 4.5\ days$$

The result is: 4.5 days are predicted for the processing time of the offer. In the present case, the basic test (g) had resulted in a processing time of 4.6 days. The mathematical model for the processing time of the offer in the form of the predictive function is therefore obviously quite reasonable.

3.7 Calculation scheme for the 2^3 factorial design

This calculation scheme is based on a fully executed 2^3 factorial design with eight runs. Table 50 shows the factor levels for the runs in standard sequence. The methodology whereby the 2^3 factorial design evolves from the 2^2 factorial design can also be seen from this Table.

		I	A	B	AB	C	AC	BC	ABC
1	(g)	+	-	-	+	-	+	+	-
2	a	+	+	-	-	-	-	+	+
3	b	+	-	+	-	-	+	-	+
4	ab	+	+	+	+	-	-	-	-
5	c	+	-	-	+	+	-	-	+
6	ac	+	+	-	-	+	+	-	-
7	bc	+	-	+	-	+	-	+	-
8	abc	+	+	+	+	+	+	+	+

Table 50: The sign schematic for the eight runs of the full factorial 2^3 factorial design in standard sequence

The effects and the interactions are calculated using the expressions in column 1 of Table 53. The ordinates for the pairs of points in the diagrams of the effects and interactions are calculated according to Table 51 or Table 52.

Factor level	Factor A	Factor B	Factor C
-	$\frac{1}{4}((g) + b + c + bc)$	$\frac{1}{4}((g) + a + c + ac)$	$\frac{1}{4}((g) + a + b + ab)$
+	$\frac{1}{4}(a + ab + ac + abc)$	$\frac{1}{4}(b + ab + bc + abc)$	$\frac{1}{4}(c + ac + bc + abc)$

Table 51: Calculation of the pairs of points for the graphical representation of the main effects (main effects plot)

	A	B	Start and end points of line
B low	-	-	$\frac{1}{2}((g)+c)$
B low	+	-	$\frac{1}{2}(a+ac)$
B high	-	+	$\frac{1}{2}(b+bc)$
B high	+	+	$\frac{1}{2}(ab+abc)$

y=f(A) with parameter B

	A	C	Start and end points of line
C low	-	-	$\frac{1}{2}((g)+b)$
C low	+	-	$\frac{1}{2}(a+ab)$
C high	-	+	$\frac{1}{2}(c+bc)$
C high	+	+	$\frac{1}{2}(ac+abc)$

y=f(A) with parameter C

	B	C	Start and end points of line
C low	-	-	$\frac{1}{2}((g)+a)$
C low	+	-	$\frac{1}{2}(b+ab)$
C high	-	+	$\frac{1}{2}(c+ac)$
C high	+	+	$\frac{1}{2}(bc+abc)$

y=f(B) with parameter C

Table 52: Start and end points for the diagrams of the interactions AB, AC and BC

Effects	Coefficients $b_0 = \bar{y}$	$MS_I = SS_I$	$F = \dfrac{MS_I}{MS_R}$	Significant if
$A = \dfrac{1}{4}(a + ab + ac + abc)$ $-\dfrac{1}{4}((g) + b + c + bc)$	$b_1 = \dfrac{A}{2}$	$2A^2$	$F_A = \dfrac{2A^2}{MS_R}$	$F_A > F_{1-\alpha}(1; f_R)$
$B = \dfrac{1}{4}(b + ab + bc + abc)$ $-\dfrac{1}{4}((g) + a + c + ac)$	$b_2 = \dfrac{B}{2}$	$2B^2$	$F_B = \dfrac{2B^2}{MS_R}$	$F_B > F_{1-\alpha}(1; f_R)$
$AB = \dfrac{1}{4}((g) + ab + c + abc)$ $-\dfrac{1}{4}(a + b + ac + bc)$	$b_3 = \dfrac{AB}{2}$	$2(AB)^2$	$F_{AB} = \dfrac{2(AB)^2}{MS_R}$	$F_{AB} > F_{1-\alpha}(1; f_R)$
$C = \dfrac{1}{4}(c + ac + bc + abc)$ $-\dfrac{1}{4}((g) + a + b + ab)$	$b_4 = \dfrac{C}{2}$	$2C^2$	$F_C = \dfrac{2C^2}{MS_R}$	$F_C > F_{1-\alpha}(1; f_R)$
$AC = \dfrac{1}{4}((g) + b + ac + abc)$ $-\dfrac{1}{4}(a + ab + c + bc)$	$b_5 = \dfrac{AC}{2}$	$2(AC)^2$	$F_{AC} = \dfrac{2(AC)^2}{MS_R}$	$F_{AC} > F_{1-\alpha}(1; f_R)$
$BC = \dfrac{1}{4}((g) + a + bc + abc)$ $-\dfrac{1}{4}(b + ab + c + ac)$	$b_6 = \dfrac{BC}{2}$	$2(BC)^2$	$F_{BC} = \dfrac{2(BC)^2}{MS_R}$	$F_{BC} > F_{1-\alpha}(1; f_R)$
$ABC = \dfrac{1}{4}(a + b + c + abc)$ $-\dfrac{1}{4}((g) + ab + ac + bc)$	$b_7 = \dfrac{ABC}{2}$	$2(ABC)^2$	$F_{ABC} = \dfrac{2(ABC)^2}{MS_R}$	$F_{ABC} > F_{1-\alpha}(1; f_R)$

Table 53: Calculation scheme with Analysis of Variance for the 2^3 factorial design.
(The significance of the effects can also be checked using the p-value:
They are significant when $p < \alpha$).

The sums of squares SS_R (as a measure of the variance/error estimation) are either known or must be obtained from the observations of the effects which are not significant. The 2^3 factorial design has a 3-factor interaction ABC, which is rarely significant. This can therefore be used together with further not significant 2-factor interactions to estimate the variance (pooling).

The mean squares are calculated in accordance with:

$$MS_R = \frac{SS_R}{f_R}$$

We need to specify the confidence level $1 - \alpha$ (usually 95 % or 90 %) with which the test is to be performed. Note that how meaningful the F-test is depends strongly on the degree of freedom f_R. For factorial designs executed only once (without repetitions) with only two or three factors and thus only few degrees of freedom to estimate the error, the statistical basis of the significance test is possibly insufficient.

4th step:	Draw up predictive function

$$y = \bar{y} + \frac{A}{2}x_A + \frac{B}{2}x_B + \frac{AB}{2}x_Ax_B + \frac{C}{2}x_C + \frac{AC}{2}x_Ax_C + \frac{BC}{2}x_Bx_C + \frac{ABC}{2}x_Ax_Bx_C$$

or

$$y = b_0 + b_1x_A + b_2x_B + b_3x_Ax_B + b_4x_C + b_5x_Ax_C + b_6x_Bx_C + b_7x_Ax_Bx_C$$

The factors $b_1, b_2, ... b_7$ corresponding to not significant effects are set to zero in these functional equations.

The following normalisation is carried out for the independent variables x_A to x_C:

$$x_A = \frac{A^* - \frac{1}{2}(A_2 + A_1)}{\frac{1}{2}(A_2 - A_1)} \qquad x_B = \frac{B^* - \frac{1}{2}(B_2 + B_1)}{\frac{1}{2}(B_2 - B_1)} \qquad x_C = \frac{C^* - \frac{1}{2}(C_2 + C_1)}{\frac{1}{2}(C_2 - C_1)}$$

x_A to x_C are the minimum or maximum normalised values which each lie in the range -1 to +1.
A_1 to C_1 and A_2 to C_2 are the physical values of the low and high level of the factors.
A^* to C^* are the factor settings for which the response y to be expected is to be calculated.

4 Factorial Designs With More Than Three Factors (Systematic Structure of 2^k Designs)

In the previous chapters, 2^2 and 2^3 factorial designs have been developed with the aid of examples and the analogous step-by-step procedure has been explained. The findings thus gained are now to be extended to a general 2^k factorial design.

The runs are shown here in the familiar standard sequence. This does not mean that the runs should be carried out in this sequence, however. On the contrary: the sequence of the runs should be random so that unknown, uncontrollable variables can cancel each other out. This strategy, which is called randomising, is explained in more detail in Chapter 6 with the aid of an example. Until then, we will make do without randomising for reasons of clarity.

The rules for the signs of the observations to compute the effects and interactions are as follows:

Main Effects   (Rule 1)

Observations whose designations contain the corresponding factor name are input into the computation of the effects and interactions with a positive sign, the others with a negative one.

2-factor interactions   (Rule 2)

Observations whose designations contain either both corresponding factor names or none of the corresponding factor names are input with a positive sign, the others with a negative one.

3-factor interactions   (Rule 3)

Observations whose designations contain either one or all three corresponding factor names are input into the computation of the effects and interactions with a positive sign, the others with a negative one.

Interactions between more than three factors are hardly significant. If the corresponding runs are carried out, the observations are useful to estimate the experimental error, as has been shown. The sign to compute multiple interactions can be seen from the sign schematic (Table 54), which is explained in more detail below.

In the transition from the 2^2 to the 2^3 factorial design, four new runs c, ac, bc and abc have been added because of the additional factor C. The associated signs for the effects A, B, and AB are obtained as per Rule 1. They are identical to those of the runs (g), a, b and ab. The signs for the effect C also result from Rule 1, those of the interactions AC, BC and ABC from Rule 2 or 3, or simply from the signs of the main effects by applying the multiplication rule.

Table 54 contains the signs for the computation of all effects and interactions for 2^2, 2^3 and 2^4 factorial designs. The systematic repetition of complete "blocks of signs" when moving to the next higher factorial design can be seen: the block formed by $I/(g)$ to AB/ab is repeated three times (starting at I/c, at I/d and at I/cd), for example. A necessary condition for the plausibility of the matrix is that the sums of the columns (except in column I) are all equal to zero.

			2^2				2^3							2^4	
	I	A	B	AB	C	AC	BC	ABC	D	AD	BD	ABD	CD	ACD	BCD	ABCD
(g)	+	-	-	+	-	+	+	-	-	+	+	-	+	-	-	+
a	+	+	-	-	-	-	+	+	-	-	+	+	+	+	-	-
b	+	-	+	-	-	+	-	+	-	+	-	+	+	-	+	-
ab	+	+	+	+	-	-	-	-	-	-	-	-	+	+	+	+
c	+	-	-	+	+	-	-	+	-	+	+	-	-	+	+	-
ac	+	+	-	-	+	+	-	-	-	-	+	+	-	-	+	+
bc	+	-	+	-	+	-	+	-	-	+	-	+	-	+	-	+
abc	+	+	+	+	+	+	+	+	-	-	-	-	-	-	-	-
d	+	-	-	+	-	+	+	-	+	-	-	+	-	+	+	-
ad	+	+	-	-	-	-	+	+	+	+	-	-	-	-	+	+
bd	+	-	+	-	-	+	-	+	+	-	+	-	-	+	-	+
abd	+	+	+	+	-	-	-	-	+	+	+	+	-	-	-	-
cd	+	-	-	+	+	-	-	+	+	-	-	+	+	-	-	+
acd	+	+	-	-	+	+	-	-	+	+	-	-	+	+	-	-
bcd	+	-	+	-	+	-	+	-	+	-	+	-	+	-	+	-
abcd	+	+	+	+	+	+	+	+	+	+	+	+	+	+	+	+

Table 54: The sign schematic to compute the effects for 2^2, 2^3 and 2^4 factorial designs

There is therefore a systematic repetition of combinations of signs block-by-block.
Table 55 shows these up to the 2^4 factorial design. It can be expanded to 2^k factorial designs using the following rules:

- If the block contains a negative sign, all signs of the effects in this block have the opposite sign to the block on the immediate left.

- If the block contains a positive sign, all signs are identical to those in the block on the immediate left.

The identity column must be included in the block consideration in each case.

The effects and interactions are computed by giving the observations the sign of the corresponding columns, adding them together and averaging them. Table 56 shows this using the example of effect A.

	I	A	B	AB	C	AC	BC	ABC	D	AD	BD	ABD	CD	ACD	BCD	ABCD
(g)	1	-1		–				–				–				
a	1	1														
b		+		+												
ab																
c						+										
ac		+														
bc																
abc																
d																
ad																
bd																
abd				+									+			
cd																
acd																
bcd																
abcd																

Table 55: The systematic feature of the sign schematic, expandable to 2^k factorial designs. Block with -/+: flip or take over the sign of the block to the immediate left.

Effect A	$MS_A = SS_A$	Design
$A = \frac{1}{2}(a + ab)) - \frac{1}{2}((g) + b)$	A^2	2^2
$A = \frac{1}{4}(a + ab + ac + abc) - \frac{1}{4}((g) + b + c + bc)$	$2A^2$	2^3
$A = \frac{1}{8}(a + ab + ac + abc + ad + abd + acd + abcd)$ $- \frac{1}{8}((g) + b + c + bc + d + bd + cd + bcd)$	$4A^2$	2^4

Table 56: Computation of the effects and mean squares using the factor A as an example for 2^2, 2^3 and 2^4 factorial designs

The further steps to assess the significance of the effects and interactions and the creation of the predictive function are done by analogy with the procedure for 2^2 and 2^3 factorial designs. They will be explained in detail in the next chapter with the aid of a numerical example.

This chapter will show only the systematic structure of 2^k factorial designs. On the practical level, assistance is provided by DoE tools which are based on this systematic structure and the arithmetic to compute the effects etc.

4.1 A 2^4 factorial design using the "Distillate concentration" example

The procedure for a fully executed 2^4 factorial design and the necessary equations are explained step-by-step in this chapter using a numerical example. The general calculation schematic is then presented in Chapter 4.3.

Task: Distillate concentration

A mixture of two substances is separated in a rectifying column[18]. The response is the distillate concentration which is to be investigated with the aid of a 2^4 factorial design.
Four input variables each on two levels (cf. Table 57) are under investigation[19].

	Response: Concentration [%]			Factor levels	
	Factor	Measurement unit	Factor type	Low	High
A	Reflux ratio	-	Quantitative	3	4
B	Evaporator capacity	kcal/h	Quantitative	100 000	120 000
C	Feed concentration	%	Quantitative	30	40
D	Coolant flow rate	kg/h	Quantitative	350	400

Table 57: The factors and their levels for the 2^4 factorial design "Distillate concentration"

1st step:	Plan and conduct experiments

In the present case of a fully executed 2^4 factorial design, distillate concentrations were determined for the 16 runs as per Table 58.

---

[18] In the rectification, (also called distillation in the simplest case) the substances contained in a mixture of substances are separated by heating since they have different boiling points. In large scale industrial plants, this is done in columns with rectification trays to extract the distillates at different heights.

[19] Following Engelmann, H.-D., Erdmann H.-H., Simmrock, K.H.: Planen und Auswerten von Versuchen

	Concentration [%]	Effects		b_0 to b_{15}	$SS_I = MS_I$
(g)	84.15			81.55	
a	82.00	A	3.20	1.60	40.96
b	75.50	B	-2.94	-1.47	34.52
ab	80.60	AB	2.04	1.02	16.61
c	83.70	C	2.40	1.20	23.04
ac	85.50	AC	1.55	0.78	9.61
bc	80.20	BC	0.84	0.42	2.81
abc	85.20	ABC	-0.49	-0.24	0.95
d	81.20	D	-1.11	-0.56	4.95
ad	81.60	AD	0.76	0.38	2.33
bd	77.25	BD	0.53	0.26	1.10
abd	80.50	ABD	-0.57	-0.29	1.32
cd	80.70	CD	-0.69	-0.34	1.89
acd	85.30	ACD	0.59	0.29	1.38
bcd	76.90	BCD	-0.73	-0.36	2.10
abcd	84.50	ABCD	0.52	0.26	1.10

Table 58: Observations, effects and coefficients of the predictive function of the 2^4 factorial design "Distillate concentration"

2nd step: Calculation and graphical representation of the effects and interactions

The sign schematic helps with the computation of the effects and interactions: The 16 observations are each given the sign of the corresponding columns of Table 54, added together and averaged.

Table 58 additionally shows the coefficients of the predictive function. These correspond to the effects divided by two. The last column contains the sums of squares as four times the squares of the effects.

The *main effects plots* and the *interaction plots* serve to provide a graphical illustration of the effects/interactions. Figure 36 shows the plots for the main effects and the relatively strong *AB* interaction.

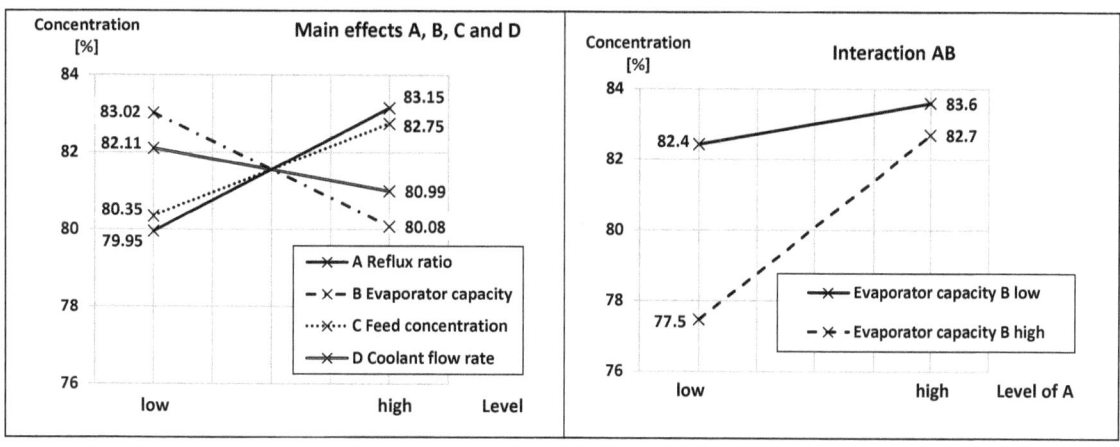

Figure 36: Main effects and AB interaction of the 2^4 factorial design "Distillate concentration"

3rd step: Test the significance of the effects and interactions (Analysis of Variance with F-test)

In the following, the (probably) not significant multiple interactions are used step-by-step to estimate the experimental error SS_R .

We start with the interactions between three and four factors, which experience has shown are not expected to be significant[20].

$$SS_R = SS_{ABC} + SS_{ABD} + SS_{ACD} + SS_{BCD} + SS_{ABCD} = 0.95 + 1.32 + 1.38 + 2.10 + 1.10 = 6.85$$

The degree of freedom is: $f_R = f_{ABC} + f_{ABD} + f_{ACD} + f_{BCD} + f_{ABCD} = 1 + 1 + 1 + 1 + 1 = 5$

The mean square is thus: $MS_R = \frac{SS_R}{f_R} = \frac{6.85}{5} = 1.37$

This produces the F-values according to Table 59, which must be compared with the critical value of the F-distribution $F_{0.95}(1; 5) \approx 6.6$.

---

[20] Rounded values are used in the computations below for reasons of clarity. This also applies to the representation of the values in the tables.

Effects		SS$_I$=MS$_I$	F=MS$_I$/MS$_R$	p-value
A	3.20	40.96	29.86	0.003
B	-2.94	34.52	25.16	0.004
AB	2.04	16.61	12.11	0.018
C	2.40	23.04	16.80	0.009
AC	1.55	9.61	7.01	0.046
BC	0.84	2.81	2.05	0.212
D	-1.11	4.95	3.61	0.116
AD	0.76	2.33	1.70	0.250
BD	0.53	1.10	0.80	0.411
CD	-0.69	1.89	1.38	0.293

Table 59: F-values to test the significance of the effects; the 3-fold and 4-fold interactions were used for the error estimation

It can be seen that the 2-fold interactions *BC, AD, BD* and *CD* will lead to the result "not significant" when compared to the critical value of the distribution. They are therefore used in the next step by further pooling in addition to the multiple interactions for better estimation of the error.

$$SS_R \approx 6.85 + SS_{BC} + SS_{AD} + SS_{BD} + SS_{CD} = 6.85 + 2.81 + 2.33 + 1.10 + 1.89 = 14.98$$

The degree of freedom is: $f_R = 5 + f_{BC} + f_{AD} + f_{BD} + f_{CD} = 5 + 1 + 1 + 1 + 1 = 9$

The mean square is thus: $MS_R = \frac{SS_R}{f_R} = \frac{14.98}{9} \approx 1.66$

This results in changed F-values for the effects as shown in Table 60. They must now be compared with the critical value of the F-distribution $F_{0.95}(1; 9) \approx 5.12$.

Effects		SS$_I$=MS$_I$	F=MS$_I$/MS$_R$	p-value
A	3.20	40.96	24.60	0.001
B	-2.94	34.52	20.73	0.001
AB	2.04	16.61	9.97	0.012
C	2.40	23.04	13.84	0.005
AC	1.55	9.61	5.77	0.040
D	-1.11	4.95	2.97	0.145

Table 60: F-values to test the significance of the effects; further 2-fold interactions were used for the error estimation

The F-value of the main effect D leads to the test result "not significant". SS_D is therefore included in the error estimation as well:

$$SS_R \approx 14.98 + SS_D = 14.98 + 4.95 = 19.93$$

The degree of freedom is: $f_R = 9 + f_D = 9 + 1 = 10$

The mean square is thus: $MS_R = \frac{SS_R}{f_R} = \frac{19.93}{10} \approx 1.99$

This results in the new F-values as shown in Table 61. The test results shown in the last column result from the comparison with the limit value $F_{0.95}(1; 10) \approx 4.96$.

Effects		$SS_I = MS_I$	$F = MS_I/MS_R$	p-value	F-test result
A	3.20	40.96	20.55	0.001	Significant
B	-2.94	34.52	17.32	0.002	Significant
AB	2.04	16.61	8.33	0.016	Significant
C	2.40	23.04	11.56	0.007	Significant
AC	1.55	9.61	4.82	0.053	

Table 61: Test results for the significance of the effects after incrementally improved error estimation

The result of this factorial design is then: the factors reflux ratio A, evaporator capacity B and feed concentration C have a significant effect on the response distillate concentration. In contrast, the influence of the coolant quantity D cannot be assumed to be a main effect, nor a contribution to an interaction. The significant AB interaction leads to the recognition that the reflux ratio and the evaporator capacity have a mutual influence on their respective effects. The AC interaction has narrowly "failed" the test. For a purely mathematical consideration of these data, the interaction must not be included in the predictive function. In case of doubt, its actual influence must then be assessed from what is known about the chemical and physical characteristics of the rectification process and the plant.

The general predictive function for 2^4 factorial designs is:

$$y = b_0 + b_1 x_A + b_2 x_B + b_3 x_A x_B + b_4 x_C + b_5 x_A x_C + b_6 x_B x_C + b_7 x_A x_B x_C +$$

$$b_8 x_D + b_9 x_A x_D + b_{10} x_B x_D + b_{11} x_A x_B x_D + b_{12} x_C x_D + b_{13} x_A x_C x_D + b_{14} x_B x_C x_D + b_{15} x_A x_B x_C x_D$$

For the present example and taking into account the significant effects and the *AB* interaction, the predictive function is now:

$$y = b_0 + b_1 x_A + b_2 x_B + b_3 x_A x_B + b_4 x_C = 81.55 + 1.60 x_A - 1.47 x_B + 1.02 x_A x_B + 1.20 x_C$$

The coefficients correspond to half the values of the effects, as is the case with the 2^2 and 2^3 designs, and can be seen in Table 58.

One **task** shall now be to compute the response to be expected for the following combination of factors:

Reflux ratio A: 3.8 $x_A = \dfrac{A^* - \frac{1}{2}(A_2 + A_1)}{\frac{1}{2}(A_2 - A_1)} = \dfrac{3.8 - \frac{1}{2}(4+3)}{\frac{1}{2}(4-3)} = 0.60$

Evaporator capacity B: 112 000 kcal/h $x_B = \dfrac{B^* - \frac{1}{2}(B_2 + B_1)}{\frac{1}{2}(B_2 - B_1)} = \dfrac{112\,000 - \frac{1}{2}(120\,000 + 110\,000)}{\frac{1}{2}(120\,000 - 110\,000)} = 0.20$

Feed concentration C: 34% $x_C = \dfrac{C^* - \frac{1}{2}(C_2 + C_1)}{\frac{1}{2}(C_2 - C_1)} = \dfrac{34 - \frac{1}{2}(40+30)}{\frac{1}{2}(40-30)} = -0.20.$

The response *y* is calculated as follows:

$$y = \left(81.55 + 1.60 \cdot 0.6 - 1.47 \cdot 0.2 + 1.02 \cdot 0.6 \cdot 0.2 + 1.20 \cdot (-0.2)\right)\% \approx 82.1\,\%$$

Result: A distillate concentration of 82.1 % is expected.

4.2 Multiple execution of factorial designs

The effect D and derived interactions have turned out not to be significant in the previous example of the 2^4 factorial design: the observed difference in the observations for small and large quantity of cooling water D was explained by random influences (experimental error). The runs a and ad therefore represent a double measurement, for example. In both cases, A is at high level and B and C at low level. And the same applies to all other observations where d was at high level. These observations can be assigned to the runs (g) to abc as second values in each case. The averages are then formed according to the 3rd column in Table 62. This means the original 2^4 factorial design becomes a 2^3 factorial design with repeated runs.

y_1	y_2	$\bar{y}_{12}$
(g)	d	$\overline{(g)} = \frac{(g)+d}{2}$
a	ad	$\bar{a} = \frac{a+ad}{2}$
b	bd	$\bar{b} = \frac{b+bd}{2}$
ab	abd	$\overline{ab} = \frac{ab+abd}{2}$
c	cd	$\bar{c} = \frac{c+cd}{2}$
ac	acd	$\overline{ac} = \frac{ac+acd}{2}$
bc	bcd	$\overline{bc} = \frac{bc+bcd}{2}$
abc	$abcd$	$\overline{abc} = \frac{abc+abcd}{2}$

Table 62: The superfluous observations of a 2^4 factorial design with factor D at high level (column y_2) are used for a 2^3 factorial design with repeated runs.

The 16 available observations can be assumed to be double measurements at the eight level combinations of the three effective factors: since there are more runs than with a design which is not repeated, this leads to better estimation of the experimental error, which makes the significance test (F-test) more sensitive.

Table 63 uses the numerical example of the preceding chapter to show the assignment of the observations y_2, where factor D was at a high level, to the runs (g) to abc and the thus formed averages $\bar{y}_{12}$.

	Concentration [%]		
	y_1	y_2	$\bar{y}_{12}$
(g)	84.15	81.20	82.675
a	82.00	81.60	81.800
b	75.50	77.25	76.375
ab	80.60	80.50	80.550
c	83.70	80.70	82.200
ac	85.50	85.30	85.400
bc	80.20	76.90	78.550
abc	85.20	84.50	84.850

Table 63: A 2^3 factorial design "Distillate concentration" with repeated runs (cf. Table 58)

The averages $\bar{y}_{12}$ are now used to compute the effects and interactions. For example, the following then applies to the main effect A:

$$A = A_{high} - A_{low} = \frac{1}{4}(\bar{a} + \overline{ab} + \overline{ac} + \overline{abc}) - \frac{1}{4}(\overline{(g)} + \bar{b} + \bar{c} + \overline{bc})$$

$$= \frac{1}{4}(81.8 + 80.55 + 85.4 + 84.85) - \frac{1}{4}(82.675 + 76.375 + 82.2 + 78.55) = 3.2$$

As expected, the effects and interactions computed agree with those of the original 2^4 factorial design of the preceding chapter, because the same experimental values are used in the computation.

Table 64 shows the effects and interactions and the associated sums of squares for the F-test. The mean square deviations SS_I are again calculated according to the equation

$$SS_{Effect} = \frac{N}{4}(Effect)^2$$

Note that N is the number of runs. With $N=16$, we then obtain:

$$SS_A = \frac{16}{4}A^2 = 4A^2 \qquad\qquad SS_B = 4B^2, \qquad ..., \qquad SS_{ABC} = 4(ABC)^2$$

The number of degrees of freedom per effect is again 1. This is because the difference between two measured values is formed here as well - in this case the difference between two averages.

The following equation thus applies here as well

$$MS_I = \frac{SS_I}{f_I} = SS_I$$

The values correspond to those of the 2^4 design (cf. Table 58).

Effects		$SS_I = MS_I$
A	3.20	40.96
B	-2.94	34.52
AB	2.04	16.61
C	2.40	23.04
AC	1.55	9.61
BC	0.84	2.81
ABC	-0.49	0.95

Table 64: Effects, interactions and mean square deviations of the 2^3 factorial design "Distillate concentration" with repeated runs

The experimental error is again estimated via the sum of the mean square deviations. It is of course interesting to see how large are the variances between the observations y_1 and those of the respective repeat runs y_2. In this special case with 2 values for each experiment, the following calculation must be carried out for each pair of values y_1 and y_2:

$$Square\ deviation\ of\ two\ values = \frac{1}{2}(y_1 - y_2)^2$$

Using the numerical values from Table 63, the sum of the eight mean square deviations is then given by:

$$SS_R = \frac{1}{2}[(84.15 - 81.2)^2 + (82 - 81.6)^2 + \cdots + (85.2 - 84.5)^2] \approx 16.18$$

With the eight degrees of freedom, this results in

$$MS_R = \frac{SS_R}{f_R} = \frac{16.18}{8} \approx 2.02$$

For an even better estimation of the variance, further mean squares of the order of MS_R can be included. This applies in particular to the interaction ABC, whose mean square of 0.95 is a very low value. From a statistical point of view, the interaction BC with a mean square of 2.81 would also be a possibility for estimating the variance. It is not included in the estimation in this case out of physical considerations (suspicion of a hidden effect).

The new estimation supplemented by the mean square of the effect ABC is:

$$MS_R = \frac{16.18 + 0.95}{8 + 1} \approx 1.90$$

The F-test to evaluate the significance of the effects will now be carried out with this value ($\alpha = 5\%$). The critical value of the F-distribution thus obtained is $F_{0.95}(1; 9) \approx 5.12$.

Table 65 shows that the effects A to C and the AB interaction are significant.

Effects		Coefficients	$SS_I = MS_I$	$F = MS_I / MS_R$	p-value	F-test result
		81.550				
A	3.20	1.600	40.96	21.52	0.001	Significant
B	-2.94	-1.469	34.52	18.14	0.002	Significant
AB	2.04	1.019	16.61	8.73	0.016	Significant
C	2.40	1.200	23.04	12.11	0.007	Significant
AC	1.55	0.775	9.61	5.05	0.051	
BC	0.84	0.419	2.81	1.47	0.256	

Table 65: Effects, interactions, coefficients and F-test results of the 2^3 factorial design "Distillate concentration" with repeated runs

This result is identical to that of the 2^4 factorial design in the preceding chapter. The estimation of the variance differs only slightly from the original one. The predictive function is therefore also identical to the one in the preceding chapter. The F-value of the AC interaction is just below the critical value of the F-distribution. The number-based assessment of the effect produces the result "not significant". If, for example, the F-test had been carried out with a confidence level of 90 %, however, AC would have been assessed as being "significant". Thus, when the result of the test is close, the plausibility of the results should always be checked using what is known about the process or system.

This example is intended to show how the effects of the 2^4 factorial design which were assessed as being not significant can be used in the 2^3 factorial design with repeated runs to improve the estimation of the variance.

Executing a factorial design at least twice has proven to be very successful in practice. The following applies to the n-fold repeat of factorial designs with two levels:

- The effects are calculated from the averages of the results per factor combination.

- The sums of the mean square deviations SS_I of the effects are calculated as follows:

$$SS_{Effect} = \frac{N}{4}(Effect)^2$$

where N is the total number of runs conducted.

- The respective mean square deviations of the observations associated with each factor combination are suitable to estimate the variance. They are now added together. The number of degrees of freedom for the calculation of the mean squares MS_R corresponds to the number of values used for the sum of the mean square deviations.

4.2.1 A 2^4 factorial design (repeated, randomised) using the example of the "Tear resistance of a cotton fabric"

The example explained below represents a fully executed 2^4 factorial design. Fully executed means that observations are available for all 16 level combinations. Repeating every run on every level combination will produce two runs per level combination, whose averages will be used to perform the computation, as shown in the preceding chapter.

A further, important, strategic component is introduced here in addition by not executing the 16 runs twice in the standard sequence *(g)* to *abcd*. Instead, a random sequence for the runs is imposed. This is called randomising. The great advantage of randomised factorial designs is that influences of unknown, uncontrollable variables are randomly dispersed and can thus compensate each other. If, for example, the air pressure in the pilot plant is one of these unknown, uncontrollable variables, which has a particular strong effect at one combination of levels, there is a chance of partially cancelling out this influence with the two associated runs, which are executed on different days, for example.

Example: Tear resistance of a cotton fabric

Cotton fabrics are finished using a catalyst and a cross-linker. The response investigated was the tear resistance with the aid of a 2^4 factorial design with repeated runs. The input factors were the quantity of cross-linker A, the quantity of catalyst B, the condensation temperature C and the condensation time D, which were set to the levels corresponding to Table 66[21].

	Response: Tear resistance [N]			Factor levels	
	Factor	Measurement unit	Factor type	Low	High
A	Amount of cross-linker	g/L	Quantitative	100	150
B	Quantity of catalyst	g/L	Quantitative	5	10
C	Condensation temperature	°C	Quantitative	160	170
D	Condensation time	min	Quantitative	3	5

Table 66: Input factors and their levels for the 2^4 factorial design "Tear resistance"

1st step: Plan and conduct experiments

Table 67 shows the measured values for the runs executed in randomised sequence. Table 68 shows the values in standard sequence. The respective averages $\bar{y}_{12}$ are given in the last column. They are used to compute the factorial design.

---

[21] Following Petersen, Harro: Grundlagen der Statistik und der statistischen Versuchsplanung

No.	Des.	Tear resistance [N]
1	ad	382
2	ab	344
3	b	430
4	cd	391
5	bd	374
6	b	369
7	bd	383
8	(g)	421
9	acd	349
10	bcd	325
11	a	418
12	abc	285
13	abcd	278
14	d	420
15	c	419
16	(g)	450

No.	Des.	Tear resistance [N]
17	abc	336
18	c	380
19	d	404
20	ad	381
21	acd	358
22	cd	390
23	bc	326
24	abcd	242
25	abd	300
26	ac	375
27	bc	314
28	bcd	325
29	ac	369
30	abd	270
31	a	413
32	ab	342

Table 67: The measured responses of the
2^4 *factorial design "Tear resistance" in randomised sequence with repeated runs*

	Tear resistance [N]		
	y_1	y_2	$\bar{y}_{12}$
(g)	421	450	435.5
a	418	413	415.5
b	430	369	399.5
ab	344	342	343.0
c	419	380	399.5
ac	375	369	372.0
bc	326	314	320.0
abc	285	336	310.5
d	420	404	412.0
ad	382	381	381.5
bd	374	383	378.5
abd	300	270	285.0
cd	391	390	390.5
acd	349	358	353.5
bcd	325	325	325.0
abcd	278	242	260.0

Table 68: The averages $\bar{y}_{12}$ of the measured pairs of responses for the
2^4 *factorial design "Tear resistance" in standard sequence with repeated runs*

118

The averages $\bar{y}_{12}$ serve as the computational basis for the effects, interactions etc. in Table 69. The equations for this were derived in the previous chapters. In the following chapter, they are summarised again in an overview for the 2^4 design.

It should be noted that the mean square deviations are again calculated according to the equation

$$SS_{Effect} = \frac{N}{4}(Effect)^2$$

where N is the number of runs. With N=32, this results in:

$$SS_A = \frac{32}{4}A^2 = 8A^2, \quad SS_B = 8B^2, \quad ..., \quad SS_{ABCD} = 8(ABCD)^2$$

As shown in the preceding example, the mean square deviations to estimate the variance are computed from the pairs of values y_1 and y_2. The result for this set of numerical values (Table 68) is:

$$SS_R = \frac{1}{2}[(421 - 450)^2 + (418 - 413)^2 + \cdots + (278 - 242)^2] = 5\ 754.50$$

Effects		Coefficients 361.34	$SS_I = MS_I$	$F = MS_I/MS_R$	p-value	F-test result
A	-42.44	-21.22	14 407.53	40.059	0.00001002	Significant
B	-67.31	-33.66	36 247.78	100.785	0.00000003	Significant
AB	-13.69	-6.84	1 498.78	4.167	0.05806180	
C	-39.94	-19.97	12 760.03	35.478	0.00002012	Significant
AC	7.69	3.84	472.78	1.315	0.26843270	
BC	-7.69	-3.84	472.78	1.315	0.26843270	
ABC	11.19	5.59	1 001.28	2.784	0.11465801	
D	-26.19	-13.09	5 486.28	15.254	0.00125866	Significant
AD	-14.06	-7.03	1 582.03	4.399	0.05220752	
BD	-4.94	-2.47	195.03	0.542	0.47215263	
ABD	-9.06	-4.53	657.03	1.827	0.19530314	
CD	7.94	3.97	504.03	1.401	0.25377512	
ACD	-2.19	-1.09	38.28	0.106	0.74846438	
BCD	0.44	0.22	1.53	0.004	0.94878342	
ABCD	-2.44	-1.22	47.53	0.132	0.72096427	

Table 69: Evaluation of the 2^4 factorial design "Tear resistance" (MS_R =359.66)

With the 16 degrees of freedom, the mean square deviation as a measure of the variance is given by:

$$MS_R = \frac{SS_R}{f_R} = \frac{5\,754.5}{16} \approx 359.66$$

This also allows the test statistics for the F-test to be calculated. The critical value of the F-distribution for $\alpha = 5\,\%$ is:

$$F_{0.95}(1;16) \approx 4.49$$

It can be seen from Table 69 that the main effects A, B, C and D are significant. All interactions are not significant. Their values are used in the next step to improve the estimation of the experimental error. The AD interaction, whose F-value is very close to the critical value, shall continue to stay "in the running", however.

The new average value for the sum of the mean square deviations is:

$$MS_R = \frac{5\,754.5 + SS_{AB} + SS_{AC} + SS_{BC} + SS_{ABC} + SS_{BD} + SS_{ABD} + SS_{CD} + SS_{ACD} + SS_{BCD} + SS_{ABCD}}{16 + 10} \approx$$

$$\frac{10\,643.56}{26} \approx 409.37$$

The new F-values thus result as shown in Table 70. The critical value of the F-distribution is then $F_{0.95}(1;26) \approx 4.23$.

Effects		Coefficients 361.34	$SS_I = MS_I$	$F = MS_I/MS_R$	p-value	F-test result
A	-42.44	-21.22	14 407.53	35.195	0.000002931	Significant
B	-67.31	-33.66	36 247.78	88.546	0.000000001	Significant
C	-39.94	-19.97	12 760.03	31.170	0.000007266	Significant
D	-26.19	-13.09	5 486.28	13.402	0.001124794	Significant
AD	-14.06	-7.03	1 582.03	3.865	0.060079996	

Table 70: Evaluation of the 2^4 factorial design "Tear resistance" (MS_R=409.37).

The test result for the AD interaction is: "not significant". The result of the factorial design therefore stands: the main effects A to D are significant.

4th step: Draw up predictive function

Taking into account the significant effects, the predictive function is:

$$y = b_0 + b_1 x_A + b_2 x_B + b_3 x_C + b_4 x_D = 361.34 - 21.22 x_A - 33.66 x_B - 19.97 x_C - 13.09 x_D$$

4.3 Calculation schematic for the 2^4 factorial design

This chapter presents the step-by-step planning and evaluation of a 2^4 factorial design as an overview for a full factorial design with 16 runs. In reality, DoE tools support the planning of the experiments by offering experimental variants and strategies. These DoE tools undertake all the computation and the graphical display of the results. The parameterisation of the tools requires knowledge which is to be imparted in this book, however.

1st step:	Plan and conduct experiments

The planning of an experiment is limited to the practicable specification of the factor levels. This requires knowledge of the system or process to be investigated.

The factors and their levels must be entered into the statistics software and the experiments must be carried out in the randomised sequence proposed. The observations are then input and the desired arithmetical and graphical evaluations initiated.

2nd step:	Computation and graphical representation of the effects and interactions

The computation of the effects and interactions is based on the familiar sign schematic (Table 54). The 16 observations determined are each given the sign of the corresponding columns and added together.

The main effects plots and the 2-fold interaction plots provide a graphical illustration of the effects/interactions. The computations of the start and end points of the straight line are shown in Table 71 or Table 72.

Factor level	Start and end points A	Start and end points B
-	$\frac{1}{8}((g) + b + c + bc + d + bd + cd + bcd)$	$\frac{1}{8}((g) + a + c + ac + d + ad + cd + acd)$
+	$\frac{1}{8}(a + ab + ac + abc + ad + abd + acd + abcd)$	$\frac{1}{8}(b + ab + bc + abc + bd + abd + bcd + abcd)$

Factor level	Start and end points C	Start and end points D
-	$\frac{1}{8}((g) + a + b + ab + d + ad + bd + abd)$	$\frac{1}{8}((g) + a + b + ab + c + ac + bc + abc)$
+	$\frac{1}{8}(c + ac + bc + abc + cd + acd + bcd + abcd)$	$\frac{1}{8}(d + ad + bd + abd + cd + acd + bcd + abcd)$

Table 71: Calculation of the point pairs for the graphical representation of the four main effects

	A	B	Start and end points
B low	-	-	$\frac{1}{4}((g)+c+d+cd)$
	+		$\frac{1}{4}(a+ac+ad+acd)$
B high	-	+	$\frac{1}{4}(b+bc+bd+bcd)$
	+		$\frac{1}{4}(ab+abc+abd+abcd)$
			y=f(A) with parameter B

	A	C	Start and end points
C low	-	-	$\frac{1}{4}((g)+b+d+bd)$
	+		$\frac{1}{4}(a+ab+ad+abd)$
C high	-	+	$\frac{1}{4}(c+bc+cd+bcd)$
	+		$\frac{1}{4}(ac+abc+acd+abcd)$
			y=f(A) with parameter C

	B	C	Start and end points
C low	-	-	$\frac{1}{4}((g)+a+d+ad)$
	+		$\frac{1}{4}(b+ab+bd+abd)$
C high	-	+	$\frac{1}{4}(c+ac+cd+acd)$
	+		$\frac{1}{4}(bc+abc+bcd+abcd)$
			y=f(B) with parameter C

	A	D	Start and end points
D low	-	-	$\frac{1}{4}((g)+b+c+bc)$
	+		$\frac{1}{4}(a+ab+ac+abc)$
D high	-	+	$\frac{1}{4}(d+bd+cd+bcd)$
	+		$\frac{1}{4}(ad+abd+acd+abcd)$
			y=f(A) with parameter D

	B	D	Start and end points
D low	-	-	$\frac{1}{4}((g)+a+c+ac)$
	+		$\frac{1}{4}(b+ab+bc+abc)$
D high	-	+	$\frac{1}{4}(d+ad+cd+acd)$
	+		$\frac{1}{4}(bd+abd+bcd+abcd)$
			y=f(B) with parameter D

	C	D	Start and end points
D low	-	-	$\frac{1}{4}((g)+a+b+ab)$
	+		$\frac{1}{4}(c+ac+bc+abc)$
D high	-	+	$\frac{1}{4}(d+ad+bd+abd)$
	+		$\frac{1}{4}(cd+acd+bcd+abcd)$
			y=f(C) with parameter D

*Table 72: Start and end points for the diagrams of the
2-fold interactions AB, AC, BC, AD, BD and CD*

The observations whose effects are not significant are used as a measure of the variance (error estimation). The following applies:

$$SS_I = \frac{N}{4} \cdot (Effect)^2.$$

In this case, with $N=16$ runs, the result is:

$$SS_A = \frac{16}{4} \cdot A^2, \; SS_B = 4 \cdot B^2, \ldots, SS_{ABCD} = 4 \cdot (ABCD)^2$$

In the pooling method, the sums of squares SS_I of "weak" interactions are added together to give SS_R. The experimental error is then calculated as the average of this sum:

$$MS_R = \frac{SS_R}{f_R}$$

The degree of freedom f_R here corresponds to the number of values included in the calculation of SS_R :

For the F-test, the confidence level $1 - \alpha$ with which the F-test is to be carried out must be specified. On the practical level, the test is most frequently carried out with 95 % (less frequently 90 %). If the deciding power of the F-test is too low because the estimation of the experimental error is too coarse (few degrees of freedom), runs should be repeated.

4th step: Draw up predictive function

$$y = \bar{y} + \frac{A}{2}x_A + \frac{B}{2}x_B + \frac{AB}{2}x_A x_B + \frac{C}{2}x_C + \frac{AC}{2}x_A x_C + \frac{BC}{2}x_B x_C + \frac{ABC}{2}x_A x_B x_C +$$

$$\frac{D}{2}x_D + \frac{AD}{2}x_A x_D + \frac{BD}{2}x_B x_D + \frac{ABD}{2}x_A x_B x_D + \frac{CD}{2}x_C x_D + \frac{ACD}{2}x_A x_C x_D + \frac{BCD}{2}x_B x_C x_D +$$

$$\frac{ABCD}{2}x_A x_B x_C x_D$$

or

$$y = b_0 + b_1 x_A + b_2 x_B + b_3 x_A x_B + b_4 x_C + b_5 x_A x_C + b_6 x_B x_C + b_7 x_A x_B x_C +$$

$$b_8 x_D + b_9 x_A x_D + b_{10} x_B x_D + b_{11} x_A x_B x_D + b_{12} x_C x_D + b_{13} x_A x_C x_D + b_{14} x_B x_C x_D + b_{15} x_A x_B x_C x_D$$

The factors corresponding to not significant effects $b_1, b_2, \ldots b_{15}$ are set to zero in these functional equations. The following normalisation is carried out for the independent variables x_A to x_D :

$$x_A = \frac{A^* - \frac{1}{2}(A_2 + A_1)}{\frac{1}{2}(A_2 - A_1)} \qquad x_B = \frac{B^* - \frac{1}{2}(B_2 + B_1)}{\frac{1}{2}(B_2 - B_1)} \qquad x_C = \frac{C^* - \frac{1}{2}(C_2 + C_1)}{\frac{1}{2}(C_2 - C_1)} \qquad x_D = \frac{D^* - \frac{1}{2}(D_2 + D_1)}{\frac{1}{2}(D_2 - D_1)}$$

x_A to x_D are the minimum and maximum normalised values each lying in the range -1 to +1. A_1 to D_1 and A_2 to D_2 are the physical factor values of the low and the high level.

A^* to D^* are the factor settings for which the response y to be expected is to be calculated.

Effects	Coefficients	$MS_I = SS_I$	$F = \frac{MS_I}{MS_R}$	Significant if
	$b_0 = \bar{y}$			
A	$b_1 = \frac{A}{2}$	$4A^2$	$F_A = \frac{4A^2}{MS_R}$	$F_A > F_{1-\alpha}(1; f_R)$
B	$b_2 = \frac{B}{2}$	$4B^2$	$F_B = \frac{4B^2}{MS_R}$	$F_B > F_{1-\alpha}(1; f_R)$
AB	$b_3 = \frac{AB}{2}$	$4(AB)^2$	$F_{AB} = \frac{4(AB)^2}{MS_R}$	$F_{AB} > F_{1-\alpha}(1; f_R)$
CC	$b_4 = \frac{C}{2}$	$4C^2$	$F_C = \frac{4C^2}{MS_R}$	$F_C > F_{1-\alpha}(1; f_R)$
AC	$b_5 = \frac{AC}{2}$	$4(AC)^2$	$F_{AC} = \frac{4(AC)^2}{MS_R}$	$F_{AC} > F_{1-\alpha}(1; f_R)$
BC	$b_6 = \frac{BC}{2}$	$4(BC)^2$	$F_{BC} = \frac{4(BC)^2}{MS_R}$	$F_{BC} > F_{1-\alpha}(1; f_R)$
ABC	$b_7 = \frac{ABC}{2}$	$4(ABC)^2$	$F_{ABC} = \frac{4(ABC)^2}{MS_R}$	$F_{ABC} > F_{1-\alpha}(1; f_R)$
D	$b_8 = \frac{D}{2}$	$4D^2$	$F_D = \frac{4D^2}{MS_R}$	$F_D > F_{1-\alpha}(1; f_R)$
AD	$b_9 = \frac{AD}{2}$	$4(AD)^2$	$F_{AD} = \frac{4(AD)^2}{MS_R}$	$F_{AD} > F_{1-\alpha}(1; f_R)$
BD	$b_{10} = \frac{BD}{2}$	$4(BD)^2$	$F_{BD} = \frac{4(BD)^2}{MS_R}$	$F_{BD} > F_{1-\alpha}(1; f_R)$
ABD	$b_{11} = \frac{ABD}{2}$	$4(ABD)^2$	$F_{ABD} = \frac{4(ABD)^2}{MS_R}$	$F_{ABD} > F_{1-\alpha}(1; f_R)$
CD	$b_{12} = \frac{CD}{2}$	$4(CD)^2$	$F_{CD} = \frac{4(CD)^2}{MS_R}$	$F_{CD} > F_{1-\alpha}(1; f_R)$
ACD	$b_{13} = \frac{ACD}{2}$	$4(ACD)^2$	$F_{ACD} = \frac{2(ACD)^2}{MS_R}$	$F_{ACD} > F_{1-\alpha}(1; f_R)$
BCD	$b_{14} = \frac{BCD}{2}$	$4(BCD)^2$	$F_{BCD} = \frac{4(BCD)^2}{MS_R}$	$F_{BCD} > F_{1-\alpha}(1; f_R)$
ABCD	$b_{15} = \frac{ABCD}{2}$	$4(ABCD)^2$	$F_{ABCD} = \frac{4(ABCD)^2}{MS_R}$	$F_{ABCD} > F_{1-\alpha}(1; f_R)$

Table 73: The calculation schematic with Analysis of Variance for the 2^4 factorial design

5 Fractional Factorial Designs (Methodology of 2^{k-p} Factorial Designs)

So far, only full factorial designs have been used to explain the principles of DoE. This means that all possible factor combinations have been set and the response has been measured in each case. From a mathematical point of view, this is the ideal solution, of course, because maximum information about the process is thus collected. It allows the mathematical model of the process (the predictive function) to mirror reality as well as possible. The presence of interactions is investigated. The advantage this has over other methods of designing experiments is paid for with a "large" number of runs, however, which can quickly become "unrealistic" for a correspondingly large number of factors. There can be economic reasons for it becoming unrealistic. If the experiments are not "non-destructive", like crash tests of cars, full factorial designs can strain the budgets of the test departments. The large number of runs can represent a time problem as well. Surveys forming part of sociological, medical or psychological studies often extend over longer periods. If it is several years before these factorial designs are concluded, the data may harbour the risk that not all individual runs were carried out under the same framework conditions.

As has been shown, *n-1* effects can be determined with *n* factor level combinations. The number of runs increases with increasing number of factors, however. In particular, the interactions of more than two factors force up the number of runs, as is shown in Table 74.

Design	Number of runs	Number of factors	Number of effects		
			Factors	2-factor interactions	Multiple interactions
2^2	4	2	2	1	
2^3	8	3	3	3	1
2^4	16	4	4	6	5
2^5	32	5	5	10	16
2^6	64	6	6	15	42
2^7	128	7	7	21	99
2^8	256	8	8	28	219

Table 74: For full factorial designs, the number of runs to determine possible multiple interactions increases rapidly with the number of factors

For a large number of factors, this means that a great deal of experimental effort must be expended to determine possibly significant interactions. Reality shows that multiple interactions are rarely significant, however. They can therefore be neglected. The fact that they can be used to estimate the experimental error, as explained in the earlier chapters, is an expensive side effect. So the question now is whether it is possible to do without the corresponding runs and the risks that would be incurred with this procedure.

5.1 Halving the experimental effort by not conducting specific runs

Using the example of a 2^3 full factorial design (Table 75), we will now show how the size of the factorial design can be halved by dropping certain runs. The proposal is to carry out only those runs where *ABC* has a positive sign[22]. These are the four runs for the main effects *A*, *B*, *C* and the *ABC* interaction marked with an asterisk.

		I	A	B	AB	C	AC	BC	ABC
1	(g)	+	-	-	+	-	+	+	-
*2	a	+	+	-	-	-	-	+	+
*3	b	+	-	+	-	-	+	-	+
4	ab	+	+	+	+	-	-	-	-
*5	c	+	-	-	+	+	-	-	+
6	ac	+	+	-	-	+	+	-	-
7	bc	+	-	+	-	+	-	+	-
*8	abc	+	+	+	+	+	+	+	+

Table 75: The runs in the 2^3 factorial design which "halve" the factorial design are marked with an asterisk (they are shown with + in the column ABC)

The halved factorial design with runs No. 2, 3, 5 and 8 can be illustrated with the aid of the cube model (Figure 37).

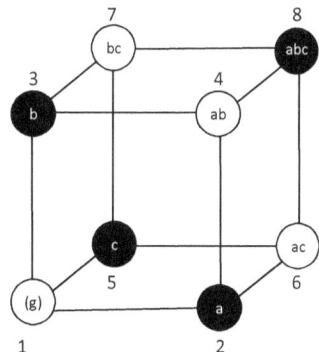

Figure 37: Halving a 2^3 factorial design: only the dark-coloured runs are carried out

---

[22] In this example, *ABC* is the so-called generator of the fractional factorial design. More details on this, for example, in: Kleppmann, Wilhelm: Taschenbuch Versuchsplanung

The halved plan then initially looks as shown in Table 76. It is called a fractional factorial design. It can now be seen that the columns AB and C are identical, for example. This means that the effect of factor C cannot be distinguished from a (possibly present) AB interaction in this reduced design. C and AB are said to be Aliases. Or: the main effect C is confounded with the AB interaction. The same applies to column B; it is confounded with AC. And A is confounded with BC. The confounding columns can also be recognised from the symmetry of the spread of the plus and minus signs in the table.

		I	A	B	AB	C	AC	BC	ABC
*2	a	+	+	-	-	-	-	+	+
*3	b	+	-	+	-	-	+	-	+
*5	c	+	-	-	+	+	-	-	+
*8	abc	+	+	+	+	+	+	+	+

Table 76: Only those runs of the full factorial design where ABC has a + are carried out for the halved 2^3 factorial design.

"Halving" has now created a fractional factorial design from the 2^3 full factorial design (Table 77). It is designated as a 2^{3-1} factorial design according to the international DoE nomenclature.

	A	B	C
a	+	-	-
b	-	+	-
c	-	-	+
abc	+	+	+

Table 77: The four runs of the 2^{3-1} factorial design in standard sequence

It should be remembered that, with the halving of the design, an eye must be kept on the following disadvantages:

- Possible AB, AC, BC interactions cannot be identified because of the confounding effects C, B or A.

- The mathematical model of the process becomes less accurate, because the effects are now computed from only four instead of the eight values previously used.

5.2 Investigate more factors with the same number of runs

The motivation in the preceding chapter was to halve the number of runs by omitting the runs to determine the interactions. These correspond to the white dots on the cube in Figure 37. We now want to demonstrate how further factors can be included in the considerations instead of the interactions while still maintaining the number of runs.

The example of the 2^2 full factorial design will be used to show how the effect of an additional factor C can be investigated for the same number of runs instead of the runs to determine the possible *AB* interaction. The top part of Table 78 shows the runs of the 2^2 full factorial design in the standard sequence. If it is assumed - or already known - that the *AB* interaction is negligibly small, run (g) could be used to determine the effect of an additional factor C as well (see line "new" in the table below). The lower part of the table now shows a 2^{3-1} factorial design. It is identical to the one in the last chapter (Table 77) which was created by halving a 2^3 full factorial design.

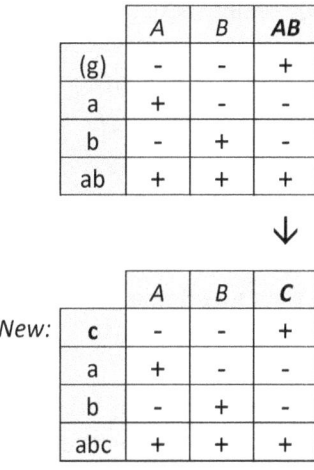

	A	B	AB
(g)	-	-	+
a	+	-	-
b	-	+	-
ab	+	+	+

↓

		A	B	C
New:	c	-	-	+
	a	+	-	-
	b	-	+	-
	abc	+	+	+

Table 78: The 2^2 full factorial design (top) becomes a 2^{3-1} fractional factorial design
(bottom table; the runs are not shown in standard sequence).

The described procedure of dropping the consideration of interactions and instead first investigating the relevance of the effects of further factors is often chosen for economic reasons. At the start of an investigation, in particular, when only little knowledge is available about a system or a machine under investigation, the primary objective is to find the significant factors from a large number of possible ones. This procedure is called screening; the associated factorial designs are called screening designs.

As has been shown, there are two equivalent ways of obtaining the same 2^{3-1} factorial design. In the first case, the factorial design is reduced to half the number of runs while maintaining the factors: in the example in the preceding chapter, a 2^3 full factorial design with eight runs became a 2^{3-1} factorial design with four runs. This chapter has shown how a 2^2 full factorial design with four runs became a 2^{3-1} fractional factorial design with four runs. These methodologies apply quite generally for the creation of fractional factorial designs.

The nomenclature for fractional factorial designs with two factor levels is as follows: 2^{k-p}

where k is the original number of factors and p the number of factors which were additionally included in the design by fractionating.

Until now, the halving of designs or the addition of *one* factor instead of the interactions has been shown for reasons of comprehensibility. For designs with more than three factors, in particular, this game of fractionating can be continued further, by taking a quarter, an eighth etc. of a design or by adding further factors. A 2^{5-1} factorial design can then be created from a 2^5 factorial design, and a 2^{5-2} can subsequently be created from this, for example. Or: a 2^5 factorial design becomes a 2^{6-1} factorial design and this becomes a 2^{7-2} factorial design and so on.

The methodology for generating 2^{k-p} factorial designs is listed in the literature in the form of tables. The term resolution of a factorial design is also used in the literature[23]. The resolution is a measure of the degree of confounding. The larger the number k of factors and the smaller the number of factor levels, the lower is the maximum achievable resolution. Vice versa: the lower the resolution, the more factors can be investigated.

The practitioner will find the logic which is explained only briefly here in the DoE software packages. After selecting a factorial design, the relevant confounded effects and interactions are displayed therein. The software tells the experimenter which interactions cannot be distinguished from the main effects.

---

[23]See also Box, George E. P. and Hunter, J. Stuart and Hunter, William G.: Statistics for Experimenters

6 Blocking and Randomising

Blocking

In many cases of actual experimental work, the framework conditions do not stay the same over the whole duration of the experiments. This can affect the observations in that (slightly) distorted values are measured. In the case that the distorting effect of the changed framework conditions is significantly smaller than the main effects, so-called blocking is a tried-and-tested method to moderate this.

This is explained with the aid of an example: a process for manufacturing a paint for exterior use is to be investigated. The response to be observed is the viscosity of the product. In order for the observations to be available as quickly as possible, the experiments are carried out on two actually identical machines. Let us assume it is known that the machines have a different impact on the viscosity; this is much smaller than the main effects, however.

It would now be wrong to carry out the runs in the standard sequence, for example, the first half being conducted on machine X and the second half on machine Y. This is because certain main effects would then be computed incorrectly, because runs with high and low levels would be conducted in unequal numbers for certain factors. Rather, the runs have to be distributed between the two machines in such a way that the machines affect the computation of the effects as little as possible. This is the case when the block variable "machine" is confounded to 100 % with a (not required) interaction. The obvious choice for 2^3 factorial designs is often the *ABC* interaction: runs where *ABC* is plus are undertaken on machine X, the others on machine Y. This ensures that the runs on both machines have the same number of plus and minus settings for all factors.

As described, fractional factorial designs have the advantage of less experimental effort. The confounded effects and interactions which thus arise produce the disadvantage that main effects and interactions cannot be distinguished, however. The blocking, in contrast, intentionally confounds the block variables responsible for the change in the framework conditions with one interaction.[24]

Randomising

In order to prevent unidentified uncontrollable variables in the process or during the measurement of the observations from distorting the overall result, the runs should not be conducted in the standard sequence, but in a random sequence. This thus creates the possibility that the uncontrollable effects cancel each other out, at least partially. The random sequence is fixed before the experiment is carried out. In reality, this task is undertaken by the random sequence generators in the DoE software. The optimum procedure is to first counteract known changes by blocking and then randomise within the blocks.

---

[24] See also Box, George E. P. and Hunter, J. Stuart and Hunter, William G.: Statistics for Experimenters or Klein, Bernd: Versuchsplanung - DoE

7 Guidelines for the Design, Execution and Evaluation of Experiments

The following procedure has proved to be successful for the planning, execution and evaluation of factorial designs in practice:

- **Clarification of task**

  - What is to be investigated and what is the objective?
  - How much time and how many resources (machines, measuring equipment, staff) are available?
  - What is known about the process?
  - Which possible factors are known?

- **Preparation of the experiments**

  - Specify factor levels
  - Specify measurement procedure and measuring equipment

- **Planning and execution of the experiments**

  - Carry out screening runs with fractional factorial designs to exclude irrelevant factors
  - Specify factorial design taking into account the objectives and economic boundary conditions (full factorial, fractional factorial)
  - Form blocks for potentially changing experimental conditions
  - Randomise the sequence of runs
  - Include star point and center point runs to identify nonlinearities.

- **Evaluation of the observations**

  - Examine completeness and plausibility of the measured results
  - Examine significance of the effects step by step with the aid of weak interactions as estimators for the experimental variance
  - Examine how well the mathematical model mirrors reality[25]
  - Submit proposals to the client regarding process improvement

---

[25] How "well" the predictive function represents the process should be determined with the aid of an analysis of residuals. The differences between the measured responses and the corresponding computed ones are compared. The differences should be similar (and similarly small) across the whole experimental space. See for example Kleppmann, Wilhelm: Taschenbuch Versuchsplanung

8 Appendix I: Literature and Internet Publications

The large number of publications in books, specialist journals and the Internet (many in English) cannot be represented in their entirety here. You will therefore find a selection of helpful publications and research possibilities below.

Books

A small selection of books is listed here which supplement the topics explained in this book in terms of content or go into more detail:

Anderson, Mark J., Whitcomb, Patrick J.: DOE Simplified–Practical Tools for Effective Experimentation, Productivity Press, New York, 2007; 2nd edition,
ISBN 978-1-56327-344-5

Anderson, Mark J., Whitcomb, Patrick J.: RSM Simplified–Optimizing Processes Using Response Surface Methods for Design of Experiments, Productivity Press, New York, 2005,
ISBN-10: 1-56327-297-0, ISBN-13: 978-1-56327-297-4

Box, George E. P. and Hunter, J. Stuart and Hunter, William G.: Statistics for Experimenters, Published by John Wiley, New York, 2nd edition 2005,
ISBN: 978-0-471-71813-0

Elser, Thomas: Statistik für die Praxis, WILEY-VCH-Verlag, Weinheim, 2004;
ISBN-3-527-50097-9

Kleppmann, Wilhelm: Taschenbuch Versuchsplanung, published by Hanser, Munich, 2008,
ISBN 978-3-446-41595-9

Klein, Bernd: Versuchsplanung - DoE, published by Oldenbourg, Munich, 2nd edition 2007,
ISBN 978-486-58352-6

Petersen, Harro: Grundlagen der Statistik und der statistischen Versuchsplanung, Volume 2, published by ecomed, Landsberg, 1991,
ISBN 3-609-65340-X

Database for research on mathematics and methodology of DoE

Helpful are also searches in the NIST database which can be accessed free of charge in order to search for mathematical or DoE technical principles. NIST is a federal operating division within the United States Department of Commerce.

NIST: National Institute of Standards and Technology

The link below takes you to the Engineering Handbook:

http://www.itl.nist.gov/div898/handbook/

Lecture scripts

Adam, Mario: Statistische Versuchsplanung und Auswertung, Hochschule Düsseldorf, Fachbereich Maschinenbau und Verfahrenstechnik, lecture script:
http://zies.hs-duesseldorf.de/Lehre/Lehrveranstaltungen/Versuchsplanung_und_Auswertung/

Ament, Ch.: Eine Einführung in die statistische Versuchsplanung, Universität Bremen, Fachgebiet Mess-, Steuerungs- und Regelungstechnik, 2002

Engelmann, H.-D., Erdmann H.-H., Simmrock, K.H.: Planen und Auswerten von Versuchen, Universität Dortmund, Fachbereich Chemietechnik, course for DECHEMA Deutsche Gesellschaft für chemisches Apparatewesen e.V., 1992

Handl, Andreas: Einführung in faktorielle und fraktionelle faktorielle Versuchspläne; lecture script for economists:
http://www.wiwi.uni-bielefeld.de/lehrbereiche/emeriti/jfrohn/skripten/Ueberblick_Handl

Jacobs, Bernhard: Einführung in die Versuchsplanung (electronic textbook), Universität des Saarlandes, Philosophische Fakultät, 1999

9 Appendix II: Examples in the Book for Downloading (MS Excel®/OpenOffice Calc®)

The numerical examples listed in the book were computed using MS Excel® and are graphically displayed. The files can also be read with OpenOffice Calc®.

The files can be downloaded free of charge from: *http://elserth.de/DoE_engl.html*.

Files for downloading

2^2 factorial designs

- Ordinal, disordinal and semi-disordinal interactions
- Product yield of a chemical reactor
- Surface roughness of turned parts
- Product yield of a chemical process (hidden effect)

2^3 factorial designs

- Product yield of a chemical reactor
- Adhesive strength of a bond
- Processing time of an offer
- Concentration of a distillate (design executed twice)

2^4 factorial designs

- Concentration of a distillate
- Tearing strength of a fabric (design executed twice)

The input fields for factor names and levels and for the measured responses etc. are highlighted in yellow.

10 Appendix III: DoE Software

The high amount of statistics software with DoE functionalities shows how widespread this method is - particularly in Anglo-American countries.

Below you will find a small selection of programs. Most of them are available in German as well as English and have test versions which can be used for a limited period of time.

- **Design Expert® (StatEase)**
  Pure DoE tool in English

  www.statease.com
  Distributor in Germany: *www.statcon.de*

- **Minitab®**
  Comprehensive statistics tool with DoE functionality, available in English and German; has become established in the Six Sigma environment
  www.additive-minitab.de

- **R**
  Free collection of statistical methods; can be adapted freely; requires a lot of detailed knowledge
  www.r-project.org

- **Statgraphics®**
  www.statgraphics.com

- **Statistica**
  www.statsoft.de

11 Appendix IV: History of Design of Experiments

The statistical experimental design methodology was developed in the 1920s by Ronald Aylmer Fisher (1890-1962). Fisher was an important theoretical biologist, geneticist, theoretical evolutionist and statistician. He originally developed the methods of experimental design and evaluation for the effective execution of agricultural trials. He investigated the effect of factors such as type, fertilisation, climate etc. on the hectare yield of agricultural products. His fundamental principles for the experimental methodology were comparability and generalisation potential, replication (the execution of an experiment several times under conditions as similar as possible), randomising (the random sequence of the runs) and blocking. Moreover, Fisher developed the Analysis of Variance method to evaluate the observations. The associated F-distribution was named for the initial of his surname.

In industry, factorial design was initially of little significance, however. This changed with the development of methods which dealt not only with the simple experimental methodology but also with optimisation problems. The statisticians George Edward Pelham Box and K.B. Wilson in particular developed these methods from the middle of the 20th century onwards. At that time, they arrived in Japan as well, where they were increasingly used in industrial development. Genichi Taguchi[26] especially integrated it – on the basis of Fisher's experiences – into a quality assurance philosophy and translated it into management language, which contributed significantly to the spread of the methodology. It has been successfully used in Japan since 1965. At the beginning of the 1980s, factorial design was increasingly associated with Japan's economic success and it was re-imported into the western world in the form as modified by Taguchi. The method has been used in the USA since 1980 and since 1985 in Germany as well. At the end of the 1980s, Dorian Shainin in the USA developed further complementary factorial design methods.

Today, Design of Experiments (DoE) has also become an important tool in the toolbox of the Six Sigma philosophy. Six Sigma is a statistical quality objective and simultaneously the name of a process improvement methodology. Its core element is the description, measurement, analysis, improvement and monitoring of business processes by statistical means. The objectives are based on parameters relating to the financial management of the company and customer needs. Six Sigma was developed in the USA in the mid-1980s by Motorola. Jack Welch successfully introduced Six Sigma at General Electric (GE) in 1996.

---

[26] See also Klein, Bernd: Versuchsplanung - DoE

Index